D1705055

DESIGN HANDBOOK
OF RESTRICTION ORIFICES

Types of Restriction Orifices:

1. One-Hole Thin Plates
2. Multi-Hole Thin Plates
3. N-Stages, Thin Plates, One-Hole, Multi-Hole
4. Long and Thick Orifices
5. Rotary Multi-Hole Thin Plates

Emilio Casado Flores
Consulting Engineer

Revision 3
August 2021

CONTENT

SCOPE OF THE HANDBOOK

The restriction orifices, ROs, are devices installed in the piping fluid systems of the power, chemical and industrial plants to reduce the pressure and limit the flow of the circulating fluid.

The scope of this handbook is to expose the design and calculation of different RO types, as are:

- The RO plates with one central hole. This RO type is the more common in all the piping fluid systems.
- The RO multi-hole plates. It is similar to the type before, with n > 3 holes instead one hole, but it has a lower cavitation coefficient.
- The multi-stage RO plates. This RO has several plates assembled in a piece of piping and every plate may have one hole or several holes. The pressure drop that it produces is high.
- The long and thick orifices. They produces a pressure drop high for low flows and may run with cavitation inside it, maintaining constant the flow rate.
- The rotary multi-hole RO plates. This RO type is new; the author has developed it as shows the technical paper published in the Hydrocarbon Processing magazine, February 2012.
  Its incipient cavitation coefficient is very close to one.
- The RO multi-hole thick plates. This RO type may run with cavitation inside it, maintaining constant flow rate.

The fluids may be liquids, saturated water, steam and gases.

CHAPTER 1

DESIGN GUIDE FOR RESTRICTION ORIFICE THIN PLATES WITH 1 CENTRAL HOLE

GUIDE Nº 1

Design guide for restriction orifice plates with 1 central hole

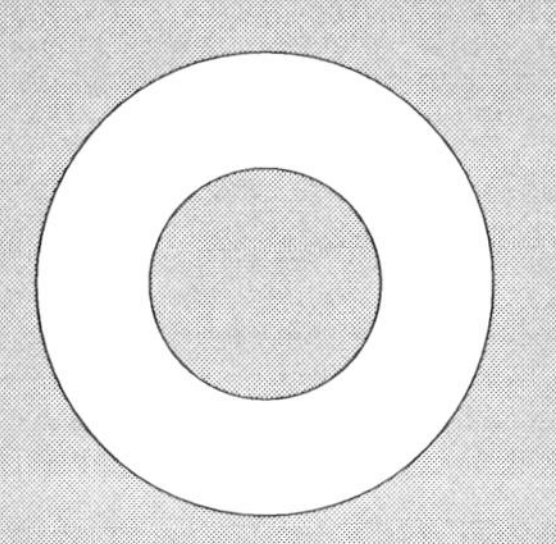

Emilio Casado

Consulting Engineer

June 2021 **Revision 2**

CONTENT

1. INTRODUCTION

The restriction orifices, named as ROs, exposed in this Guide, correspond to those formed by a circular plate with 1 central hole, as show the Figures 1 and 1a.

The hole has sharp edges and the relation between the thickness t and the hole diameter d_o usually is less than 1 and the plate is considered as a thin plate.

The ROs are the devices simpler, used in the piping systems to reduce the pressure and limit the flow.

The installation in the piping is between two flanges in a straight run, in order to recover part of the pressure downstream it.

The first elbow after the RO, in the downstream piping, must have unless a straight run of five piping diameters in order to avoid the impact of the flow with high velocity.

This revision adds the References [15] to [17] to take into account the pressure drop margins that at present, the Reference [1] doesn´t consider when the fluid through the RO is compressible.

The paragraph 3.1, also changes to be focused on saturated water.

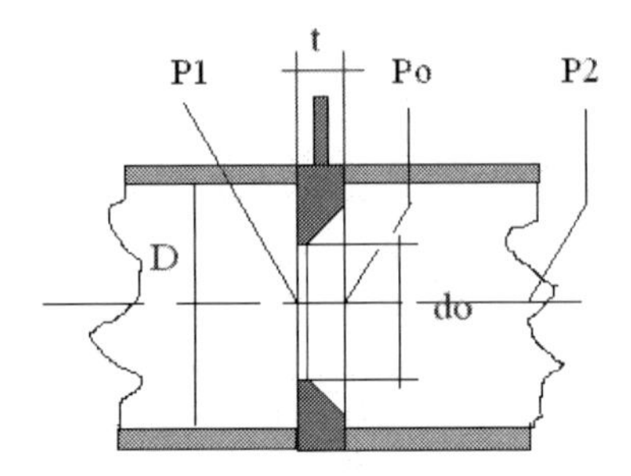

D, internal pipe diameter

do, hole diameter

t, plate thickness

Figure 1: Restriction Orifice Plate with 1 concentric hole

Figure 1a

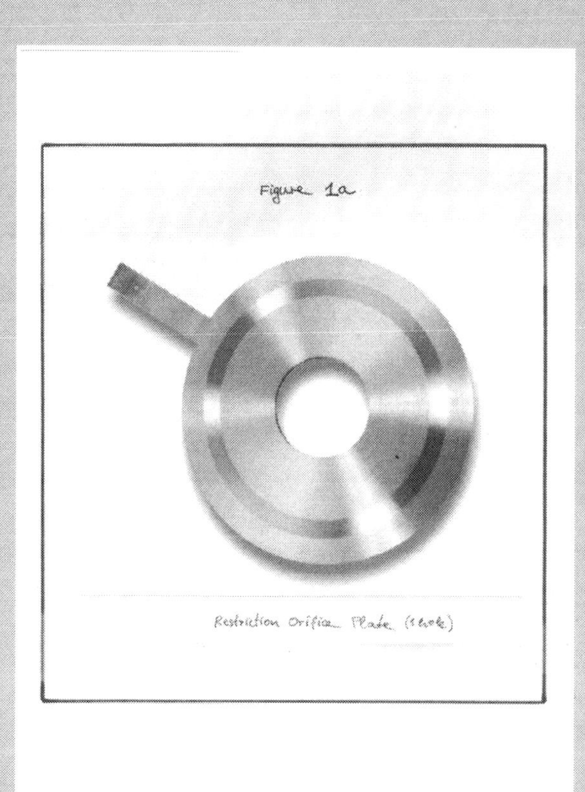

Restriction Orifice Plate (1 hole)

2. DESIGN FOR LIQUIDS

2.1 Equations that relate the flow rate and the pressure drop through the RO

The equations that relate the pressure drop, $P_1 - P_o$ and the flow rate, Q through the RO, are:

$$P_1 - P_o = 0.64Q^2/v_eC^2d_o^4 \quad (1)$$

$$Q = 1.25Cd_o^2[v_e(P_1 - P_o)]^{0,5} \quad (2)$$

The terms and their units are:

Q = Flow rate (m^3/h)

C = Flow coefficient of the RO from the Figure 2

d_o = Hole diameter (mm)

v_e = Specific volume of the fluid before the RO (m^3/kg)

P_1 = Pressure of the fluid before the RO (kg/cm^2 abs.)

P_o = Pressure at the hole outlet (kg/cm^2 abs.)

See the Reference [1].

Unless to a distance of 5D, the pressure P_2 is greater than P_o because the fluid reach the internal wall of the piping, decreasing the velocity and increasing the pressure.

The recovery pressure factor of the RO is:

$$F = (P_1 - P_2)/(P_1 - P_o) \quad (3)$$

The Figure 3 gives the values of F.

$P_1 - P_2$ is the permanent pressure loss produced by the RO.

To calculate d_o, knowing the values of $P_1 - P_2$ and Q, follow this process:

Assume a value of d_o and obtain C and F in the Figures 2 and 3. Check if accomplish the equations (1) or (2).

If not, assume other value of d_o and repeat the process until the fulfilment of the equations (1) or (2).

In the Figure 2, the Reynolds number in the axis of abscissas refers to the internal diameter of the piping before the RO and use the equation (4) to calculate it:

$$\mathbf{Re = 1273Q/mv_eD} \qquad (4)$$

The parameter m is the dynamic viscosity of the fluid in kg/m-h and D, internal diameter of the piping in mm.

After the determination of d_o, it is necessary to check that the RO has not cavitation, as the paragraph 2.2 explains.

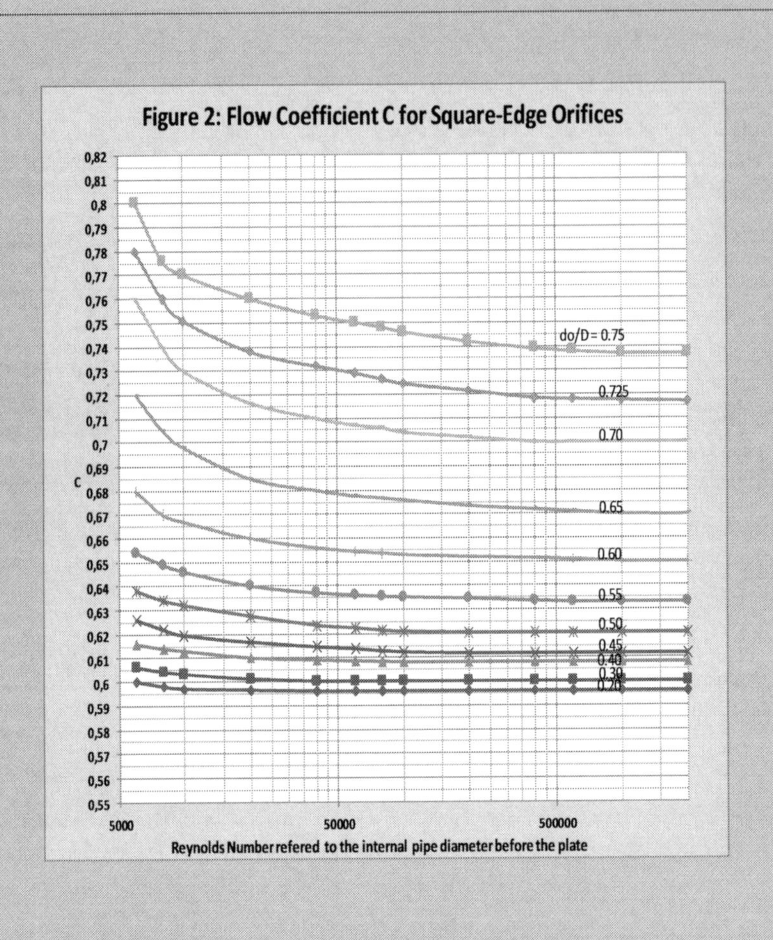
Figure 2: Flow Coefficient C for Square-Edge Orifices
0,82
0,81
0,8
0,79
0,78
0,77
0,76
0,75
0,74
0,73
0,72
0,71
0,7
0,69
0,68
0,67
0,66
0,65
0,64
0,63
0,62
0,61
0,6
0,59
0,58
0,57
0,56
0,55
C
do/D= 0.75
0.725
0.70
0.65
0.60
0.55
0.50
0.45
0.40
0.30
0.20
5000
50000
500000
Reynolds Number refered to the internal pipe diameter before the plate

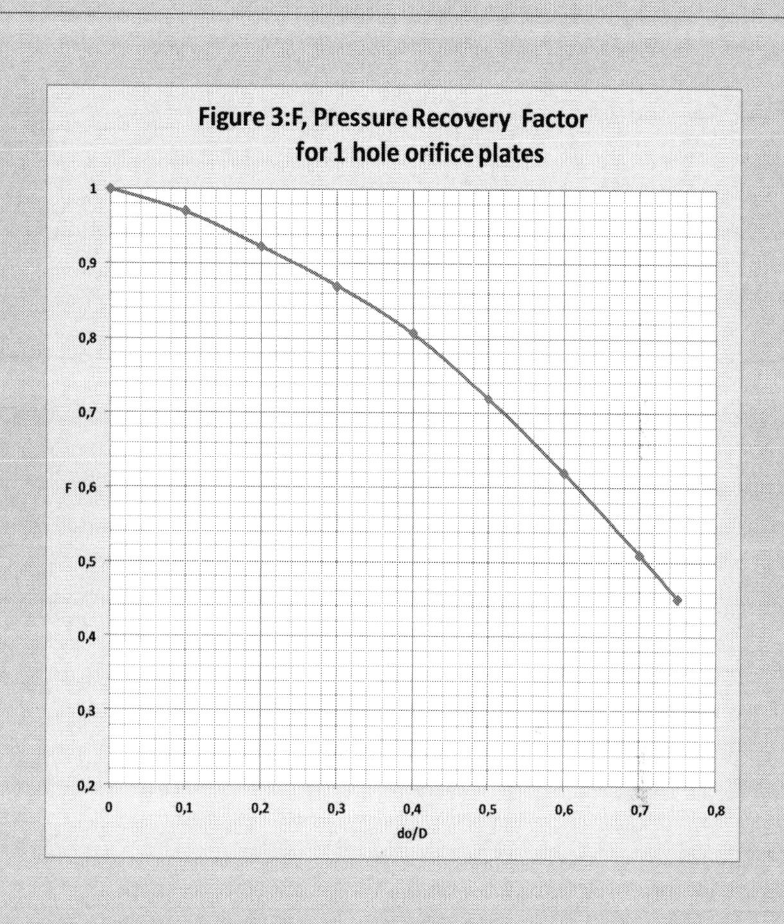
Figure 3:F, Pressure Recovery Factor
for 1 hole orifice plates
1
0,9
0,8
0,7
F 0,6
0,5
0,4
0,3
0,2
0
0,1
0,2
0,3
0,4
0,5
0,6
0,7
0,8
do/D

2.2 Cavitation

The cavitation is the implosion of the vapor bubbles that appear inside a liquid when it passes through the RO and its pressure decreases to a lower value than its vapor pressure. After the RO, the pressure increases above the vapor pressure and the vapor bubbles collapse, producing the cavitation and their consequences of noise, erosion and vibrations.

For this reason, the cavitation damage occurs in the piping downstream the RO.

Conservatively, we will consider in the design of the RO the incipient cavitation. The incipient cavitation corresponds to the beginning of the noise similar to the dragging of little stones inside the piping.

The incipient cavitation coefficient, C_i, defines the incipient cavitation. C_i is an experimental coefficient that depends on the dimensional characteristics of the RO. See in the Figure 4 the values of C_i

Note that C_i depends of the piping size D and the diameter d_o of the RO hole.

To check if the RO has cavitation, it is necessary to obtain the cavitation index applying the following equation:

$$\mathbf{s = (P_1 - P_v)/(P_1 - P_2)} \qquad (5)$$

Compare the value of s with the value of C_i. To avoid the cavitation in the RO must be $s > C_i$

P_v is the vapor pressure of the fluid.

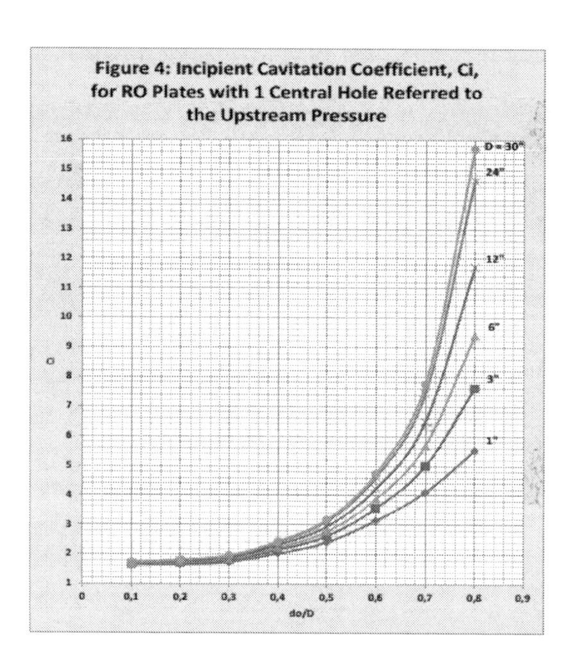
Figure 4: Incipient Cavitation Coefficient, Ci, for RO Plates with 1 Central Hole Referred to the Upstream Pressure
Ci
do/D
D = 30"
24"
12"
6"
3"
1"
16
15
14
13
12
11
10
9
8
7
6
5
4
3
2
1
0
0,1
0,2
0,3
0,4
0,5
0,6
0,7
0,8
0,9

2.3 Minimum distance between 2 places installed in series

When s is equal or less than Ci, the RO has cavitation and the solution to avoid it, is to install two ROs in series, dividing the total pressure drop $P_1 - P_2$ between the two ROs, distributed unevenly (for example 60 % in the first RO and 40 % in the second) until in every RO s be greater than Ci.

The Figure 5, taken from the Reference [7], defines the minimum distance between the two plates.

Alternatively, to the installation of two RO plates of one hole, try the design of a multi-hole plate that has lower values of C_i as shows the Reference [12].

When to avoid the cavitation were necessary to install more than two ROs in series, it´s better to design a RO of several stages as shows the Reference [13].

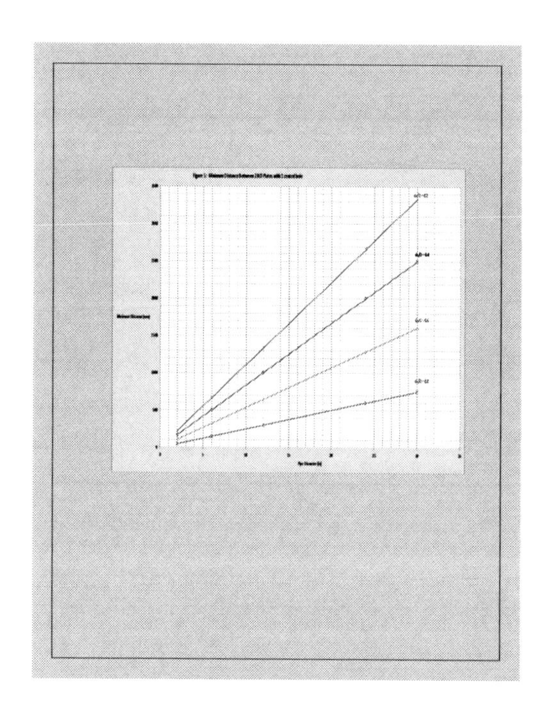

3. DESIGN FOR SATURATED WATER, STEAM AND GASES

The ROs with one sharp edge hole may have critical conditions in the pass of compressible fluids through them, at high values of the relative pressure drop $(P_1 - P_o)/P_1$

3.1 Design for saturated water

The saturated water is the fluid that flows in the drain systems of the condensate and feedwater heaters, the boilers, the steam generators and the steam pipes of the thermal and nuclear power plants. In these systems, usually are installed the RO plates.

The Reference [8] exposes the results of the tests done in RO plates with $d_o/D \leq 0.15$ and different saturation pressures $P_s = P_1$ until 15 Kg/cm^2 abs.

The Reference [15] expands the analysis of this RO plate type for saturated water to $d_o/D \leq 0.75$ and $P_s = P_1 \leq 140$ Kg/cm^2 abs.

When $P_1 = P_s$, use the equations (1) and (2) of the paragraph 2.1 for liquids, to calculate the RO plate. See the References [8] and [9].

The Reference [15], confirms that the RO plates with $d_o/D < 0.3$, practically have not critical conditions for saturated water passing through them, as also states the Reference [8], but the RO plates with $d_o/D > 0.3$, may have critical conditions.

The critical pressure $P_o = P_{CR}$ at the orifice exit is showed in the graph of the Figure 6.

Calculate the critical flow rate, applying the equation (2) with this pressure.

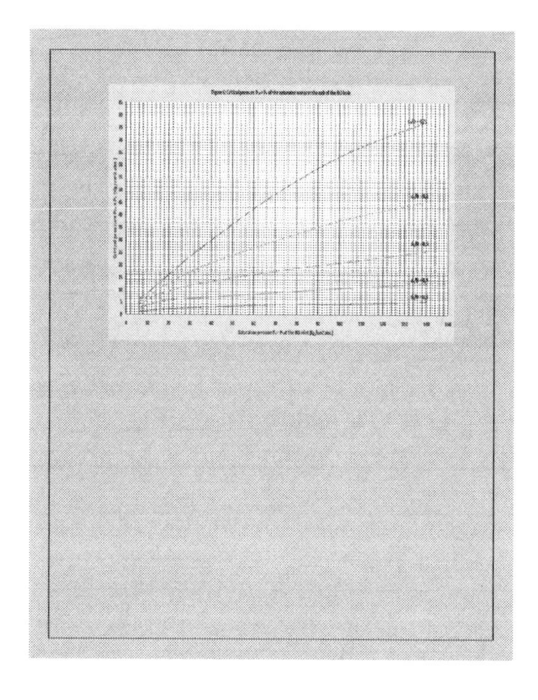

To avoid erosion damage, it is convenient to limit the fluid velocity in the RO downstream piping to a maximum of 25 m/s in continuous operation and 30 m/s in intermittent operation.

Obtain the fluid velocity in the piping with this equation:

$$\mathbf{V_2 = 3.54 v_{e2} W/D^2} \qquad (6)$$

V_2 = Fluid velocity after the RO (m/s)

W = Flow rate (kg/h)

D = Internal diameter of the piping (mm)

v_{e2} = Specific volume of the fluid after the RO (m^3/kg)

Calculate v_{e2} assuming an isentropic expansion of the saturated fluid in the RO. After the expansion from $P_1 = P_s$ to P_2, a part of the fluid flashes to vapor. Assume that $P_2 = P_o$

In the case of saturated water, the Figure 7 gives the content of steam at the RO outlet, x, in the expansion from P_s to P_2.

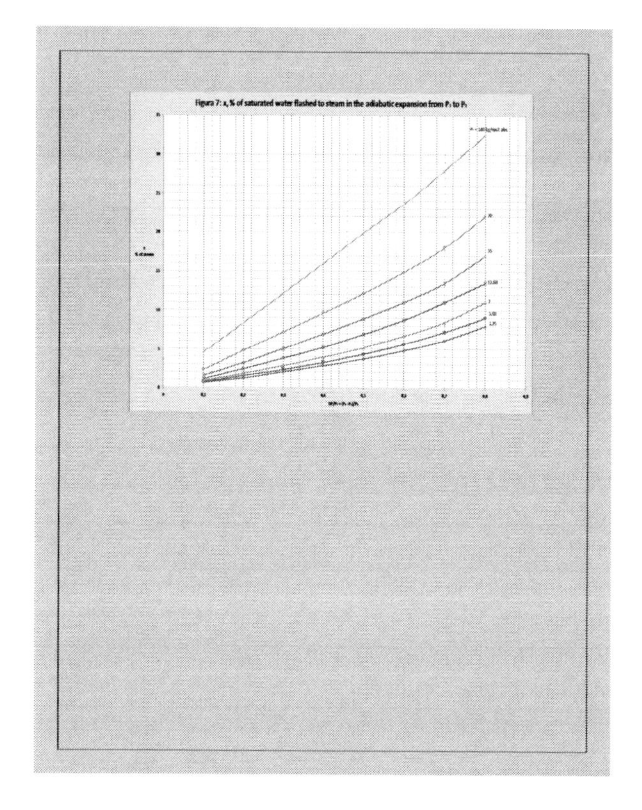

Figura 7: x, % of saturated water flashed to steam in the adiabatic expansion from P_1 to P_2

When $P_1 < P_s$ the saturated water has a vapor content $x_1 > 0$.

The References [8] and [15] show that the equations (1) and (2) change to:

$$P_1 - P_o = 0.32Q^2/v_{e1}C^2d_o^4 \quad (1a)$$

$$Q = 1.77Cd_o^2[v_{e1}(P_1 - P_o)]^{0.5} \quad (2a)$$

The critical pressure P_{CR} is the same as that of the Figure 6, multiplied by the reduction coefficient R of the Figure 8.

Calculate the value of P_s that corresponds to P_1 and x_1 using the Figure 7, trying values of P_s until find that corresponding to P_1 and x_1.

This value of P_s is used in the Figure 6 to obtain P_o and therefore will be $P_{CR} = RP_o$.

Apply also the equation (6) to limit the fluid velocity in the pipe downstream the RO.

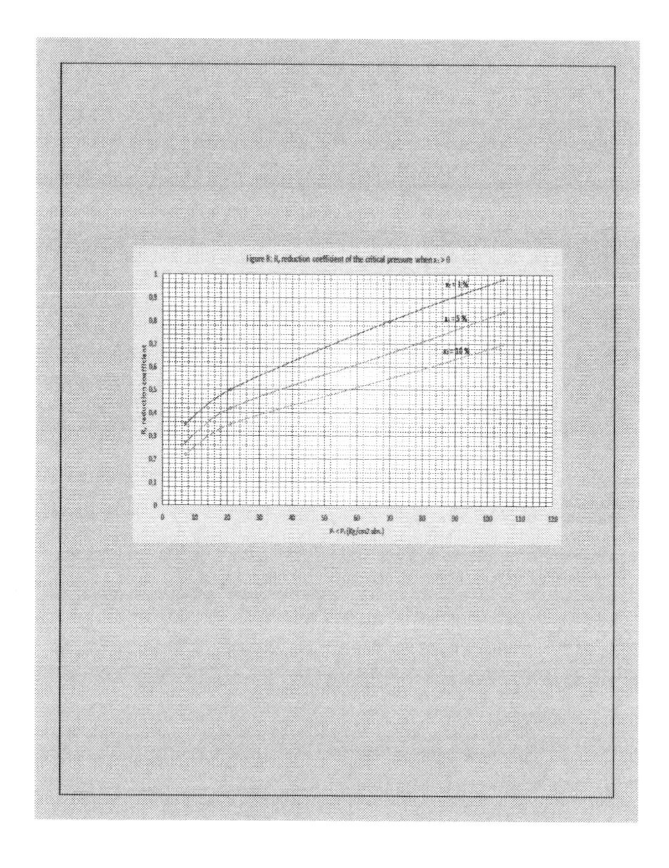

reduction coefficient of the critical pressure
1
0.9
0.8
0.7
0.6
0.5
0.4
0.3
0.2
0.1
0
0
10
20
30
40
50
60
70
80
90
100
110
120

3.2 Design for steam and gases

The equations that relate the flow rate, W in kg/h, and the pressure drop are also the equations (1) and (2) including the expansion factor Y. As $Q = v_e W$, the equations are:

$$P_1 - P_o = 0.64 v_e W^2 / Y^2 C^2 d_o^4 \quad (7)$$

$$W = 1.25 C d_o^2 Y[(P_1 - P_o)/v_e]^{0,5} \quad (8)$$

The Figures 9, 10 and 11 give the expansion factor Y for the RO.

The equation (9) gives the relation between the pressure and the specific volume before and after the RO:

$$P_1 v_{e1}^{\gamma} = P_2 v_{e2}^{\gamma} \quad (9)$$

The parameter γ is the ratio of the specific heats of the fluid at constant pressure and constant volume. Take as values of this parameter, 1.1 for saturated steam, benzene, butane, hexane, propane, and propylene; 1.3 for reheated steam, carbon dioxide, sulfur dioxide, hydrogen sulfide, ammonia, nitrogen oxide, chlorine, methane, acetylene, ethylene and natural gas; 1.4 for air, hydrogen, oxygen, nitrogen, carbon monoxide, nitric oxide and hydrochloric acid.

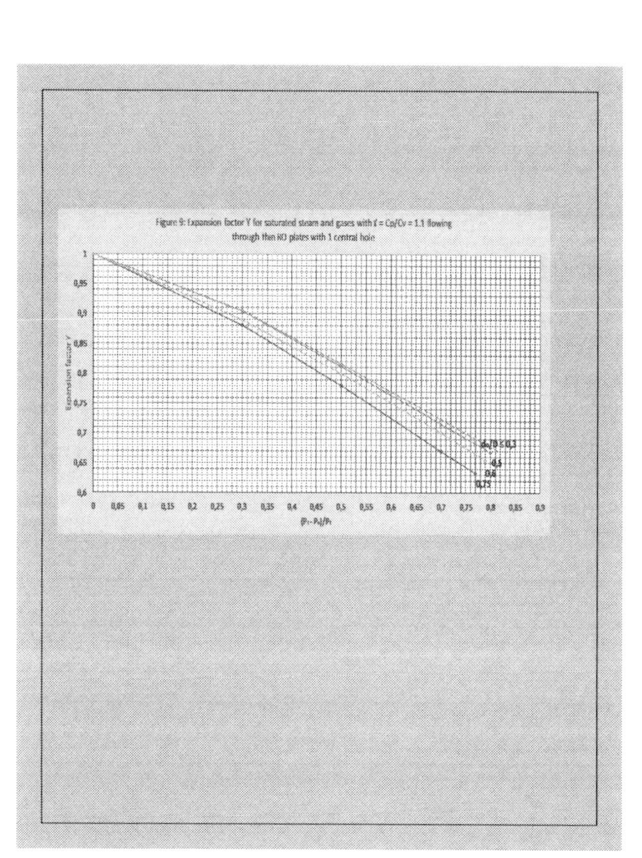

Figure 9: Expansion factor Y for saturated steam and gases with $\kappa = Cp/Cv = 1.1$ flowing through thin RO plates with 1 central hole

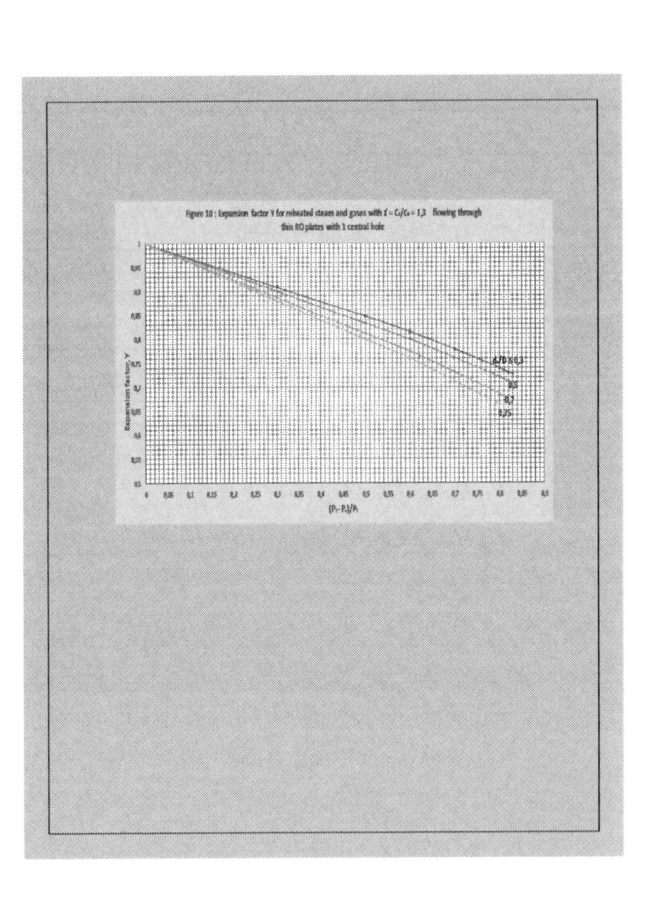
Figure 10 : Expansion factor Y for reheated steam and gases with k = Cp/Cv = 1,3 flowing through
thin RO plates with 1 central hole
Expansion factor, Y
(P1 - P2)/P1
d/D ≤ 0,3
0,5
0,7
0,75

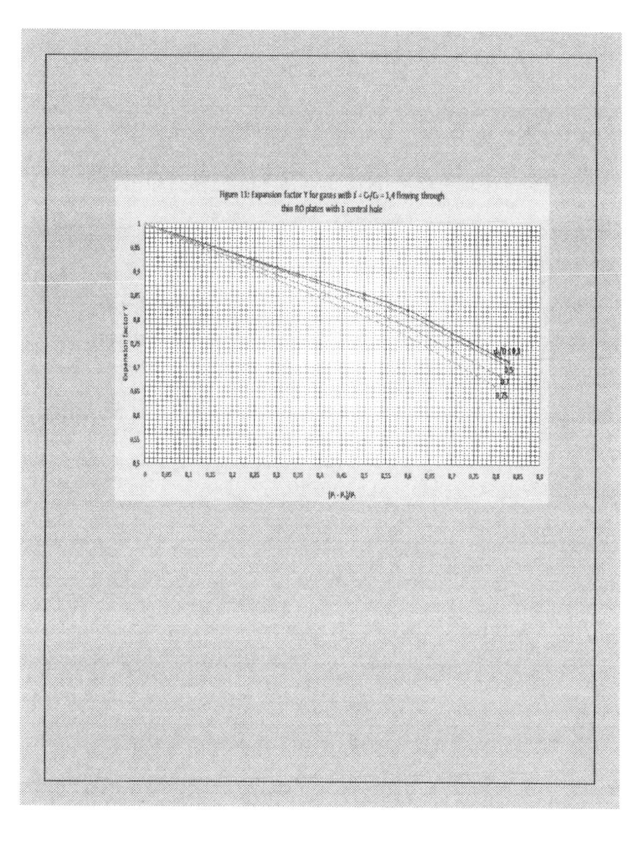
Figure 11: Expansion factor Y for gases with γ = Cp/Cv = 1,4 flowing through
thin RO plates with 1 central hole
Expansion factor Y
1
0,95
0,9
0,85
0,8
0,75
0,7
0,65
0,6
0,55
0,5
0 0,05 0,1 0,15 0,2 0,25 0,3 0,35 0,4 0,45 0,5 0,55 0,6 0,65 0,7 0,75 0,8 0,85 0,9
(P1 - P2)/P1
d/D ≤ 0,3
0,5
0,7
0,75

3.3 Noise

The RO must not provoke failures in the downstream piping, due to its acoustical energy. Use the Figure 12 to check that the RO runs in the zone without piping failures, down the Limit Curve.

Calculate the term D_2/t_2, D_2 is the internal piping diameter of the RO downstream piping and t_2 its thickness.

Calculate also the term M_2DP, M_2 is the Mach number at the RO outlet that comes from the equation (10) and $DP = P_1 - P_o$. The equation (10) is:

$$\mathbf{M_2 = 1.13(W/D_2^2)(v_{e2}/P_2g)^{0.5}} \qquad (10)$$

W = Flow rate (kg/h)

D_2 = Internal piping diameter (mm)

v_{e2} = Specific volume of the fluid after the RO (m^3/kg)

P_2 = Fluid pressure after the RO (kg/cm^2 abs.)

g = Specific heat ratio, C_P/C_V, of the fluid

Enter with D_2/t_2 in the horizontal axis of the Figure 12 and with M_2DP in the vertical axis. The crossing point is where the RO runs. If this point is in the failure zone, change the RO design, increasing t_2, installing an additional RO or installing a multi-hole RO plate.

See the References [10] y [11].

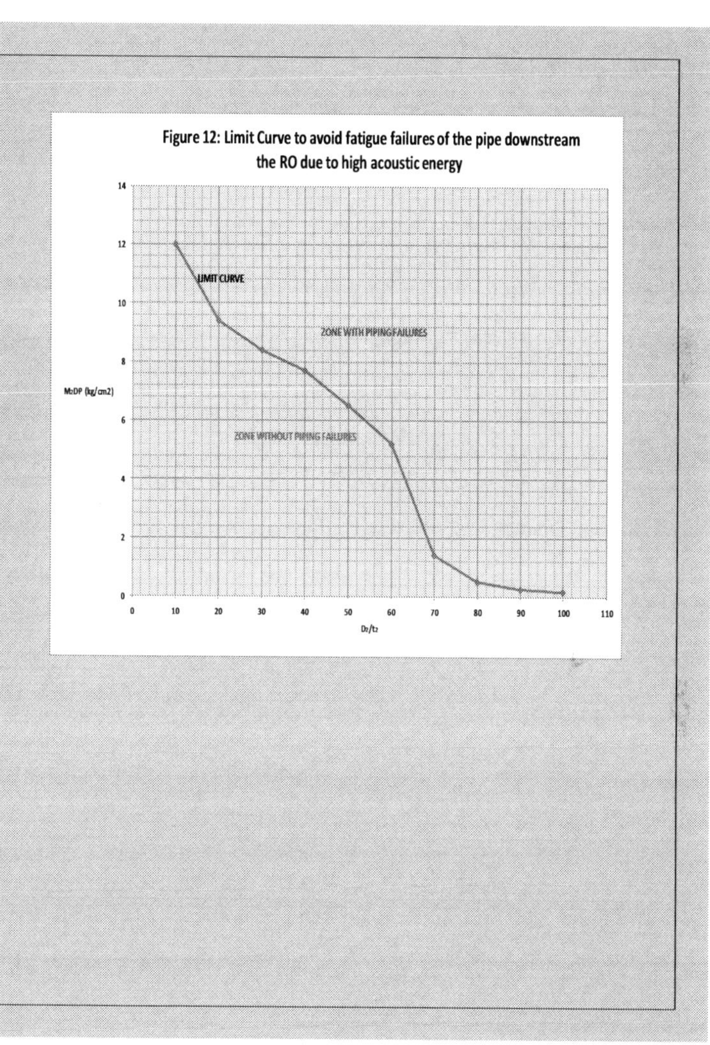

Figure 12: Limit Curve to avoid fatigue failures of the pipe downstream the RO due to high acoustic energy

4. STRUCTURAL DESIGN

The plate must have a minimum thickness to avoid its bending, due to the hydrodynamic forces of the fluid when passing through.

Use the Figure 13 to calculate directly this minimum thickness, with the internal piping diameter and the permanent pressure drop $P_1 - P_2$. This Figure is based in the Reference [14].

Alternatively, use the equation (11):

$$t = [k_2(P_1 - P_o)D^2/4S_M]^{0.5} \quad (11)$$

t = Plate thickness (mm)

$P_1 - P_o$ = Permanent pressure drop (kg/cm^2)

D = External diameter of the plate (mm)

k_2 = Correction factor taken from the Figure 14

S_M = Allowable stress of the plate material (kg/cm^2)

Use for the plate materials as stainless steel or alloyed steel

SA-240 Tp 304, 304L y 316 with S_M = 825 kg/cm^2, so in this case,

The equation (11) will be: $t = 0.0175[k_2(P_1 - P_o)D^2]^{0.5}$

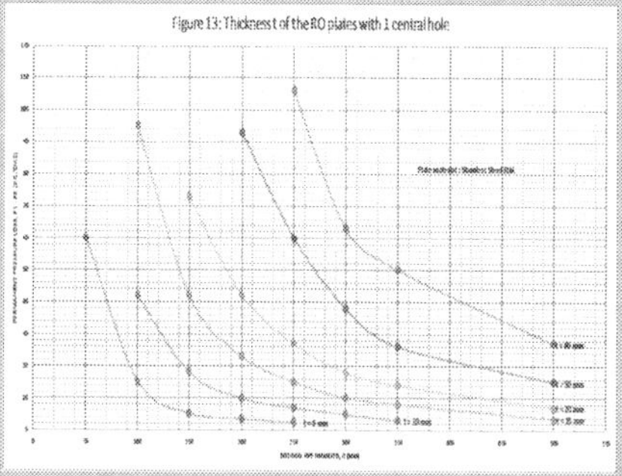

Figure 13: Thickness t of the RO plates with 1 central hole

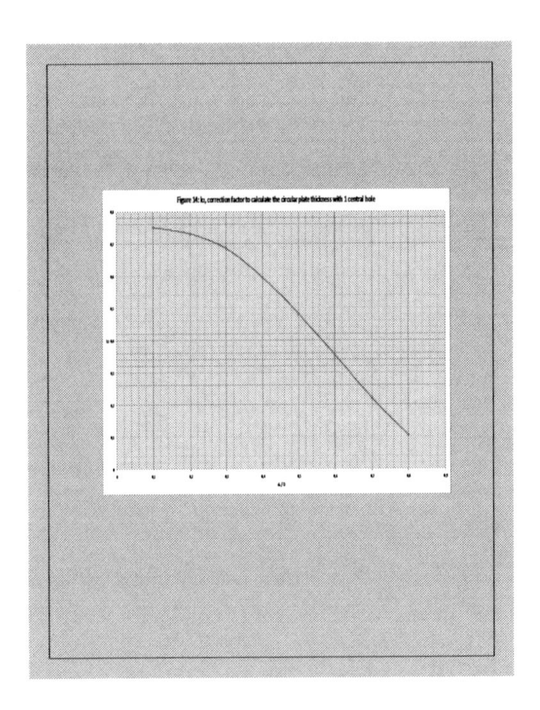

Figure 14: k₀, correction factor to calculate the circular plate thickness with 1 central hole

When the flow passing through the RO fluctuates periodically between a maximum and a minimum, it is convenient to compare the natural frequency of vibration of the plate with the excitation frequency of the flow in order to avoid the possible vibration of the plate.

The equation (12) gives the first natural frequency of vibration of a plate with a concentric hole and thickness t:

$$\mathbf{f_1 = (1.98x10^6 t/a^2)[(a^2/6 - r_o^2(b^2 - 0.5 + 0.67b^4))/ (a^2/10 - r_o^2(b^2 - 0.5 - b^4))]^{0.5}} \quad (12)$$

a = D/2

$r_o = d_o/2$

$b = r_o/a$

The term $1.98x10^6$ corresponds to $(4/p)/(gZ/rt)^{0.5}$ with g as the acceleration of the gravity, 9810 mm^2/s, Z the plate stiff in kg-mm that comes from $Et^3/10.92$, E = 21000 kg/mm^2, the Young's modulus of the steel and r = $7.8x10^{-6}$ kg/mm^3 the steel density.

The units of f_1 are cycles/s.

The second natural frequency of the plate is:

$$\mathbf{f_2 = 4.2f_1} \quad (13)$$

These equations apply when the fluid is air. For other fluids with a specific volume v_e, multiply the value of f_1 calculated with the equation (12) by the term: $1/(1 + 0.857x10^{-4}a/v_e t)^{0.5}$

If the natural frequencies are similar to the excitation frequency, increase the plate thickness to avoid the plate resonance.

See the las References [4], [5] and [6].

If the RO produces high pressure drop, it is necessary to consider the impact of the hydraulic forces in the design of the piping supports. I recommend installing an anchor or a rigid support near the RO, to avoid the piping movements induced by the passage of the fluid through the RO.

REFERENCES

[1] Crane Technical Paper No. 410 "Flow of Fluids through Valves, Fittings and Pipe"

[2] J. P. Tullis. NUREG/CR – 6031. "Cavitation Guide for Control Valves" April 1993.

[3] D. Kirk. Technical Memorandum. "Flow through Orifice Plates in Compressible Fluid Service at High Pressure Drop". December 19, 2005.

[4] W. C. Young and R. G. Budynas. "Roark's Formulas for Stress and Strain". Mc.Graw-Hill, Edition 7.

[5] S. Timoshenko. "Resistencia de Materiales". 2ª Parte. 2ª Edición.

[6] S. Timoshenko. "Vibration Problems in Engineering". Edition 2

[7] P. C. Tung and M. Mikasinovic. "Eliminating cavitation from pressure-reducing orifices". Chemical Engineering. December 12, 1983.

[8] M. W. Benjamin and J. G. Miller. "The Flow of Saturated Water Through Throttling Orifices". Transactions of the ASME. July 1941.

[9] T. J. Rholoff and I.Catton."Low Pressure Differential Discharge Characteristics of Saturated Liquids Passing Through Orifices". Transactions of the ASME. September 1996.

[10] F. L. Eisinger and J. T. Francis. "Acoustically Induced Structural Fatigue of Piping Systems" Journal of Pressure Vessel Technology. November 199. Vol. 121/438-443.

[11] F. L. Eisinger. "Designing Piping Systems Against Acoustically Induced Structural Fatigue". Journal of Pressure Vessel Technology. August 1997. Vol. 119/379-383

[12] E. Casado. "Design Guide for Multi-Hole RO Plates with n > 3 Holes. Guide Nº2: Multi-Hole Plates. February 2016.

[13] E. Casado. "Design Guide for Multi-Stage Restriction Orifice (ROs) Plates with 1 Hole or n > 3 Holes. Guide Nº 3: ROs with

N > 1 Stages. March 2016.

[14] Rototherm booklet

[15] E. Casado. "The flow of saturated water through restriction orifice thin plates". Revision 3. Research Gate, June 2021.

[16] R. G. Cunningham. "Orifice meters with supercritical compressible flow". TRANSACTIONS OF THE ASME. July 1951, pp 625-638.

[17] T. Stanton. "The flow of gases at high speeds". Proceedings of the Royal Society. Series A, Vol. 111, p-306. 1926.

CHAPTER 2

DESIGN GUIDE FOR MULTI-HOLE RO PLATES WITH MORE THAN 3 HOLES

Design Guide for Multi-Hole RO Plates

with n > 3 Holes

Emilio Casado Flores

July 2021

Revision 3

CONTENT

1. INTRODUCTION

The Restriction Orifice (RO) plates with more than 3 holes, known as multi-hole plates, have as the main advantage over the RO plates with only 1 central hole, that the cavitation coefficients are 3 or 4 times lower for the same flow rate and pressure drop.

The holes are sharp-edged and when the fluid is saturated water, steam or a gas, the RO may have critical conditions at relative high pressure drops. Likewise, the relation between the plate thickness and the equivalent diameter of the holes, t/d_e, must be less than 1 to consider the plate as a thin plate.

This Guide shows the hydraulic and structural design of this type of RO, giving the equations that relate the flow rate, the pressure drop and the required dimensions.

2. TEST PERFORMED WITH MULTI-HOLE PLATES

This type of RO plates has been tested by the author in 1991 in the Spanish nuclear power plant of Almaraz using as testing system the minimum flow recirculation line of the diesel fire pump. The results were published in the References [1] and [2].

The tested plates had n = 21 holes and pipe sizes D of 4 and 8 inches and de/D = 0.29, 0.5 and 0.68, being de the equivalent diameter of the holes, calculated with the following equation:

$$d_e = d_o n^{0.5} \qquad (1)$$

The diameter of each hole is d_o and n, the number of holes. The equivalent diameter d_e is the diameter of the area of all the holes, given by $\pi n d_o^2/4$.

The instrumentation of the tests was calibrated flow meters and manometers with an accuracy of +/- 0.1 kg/cm^2 and a range of 0 – 10 kg/cm^2 and more than 150 tests were performed.

In the next pages there are photographs of the installation for the tests with the tested plates and other different multi-hole plates designed and installed by the author in several nuclear and thermal power plants.

The number of holes usually is 13 or 21, but for pipe sizes equal or less 4 in, the number of holes could be 5 or 9.

Since these tests, a great number of different multi-hole plates with sizes from D = 3 in to D = 30 in and n = 5, 9, 13 and 21 holes has been installed in different thermal and nuclear power plants as Thiva CCP, Almaraz, Trillo, J. Cabrera and Flammanville NPP, with very good results.

See the Reference [3] that shows the experience with this type of RO.

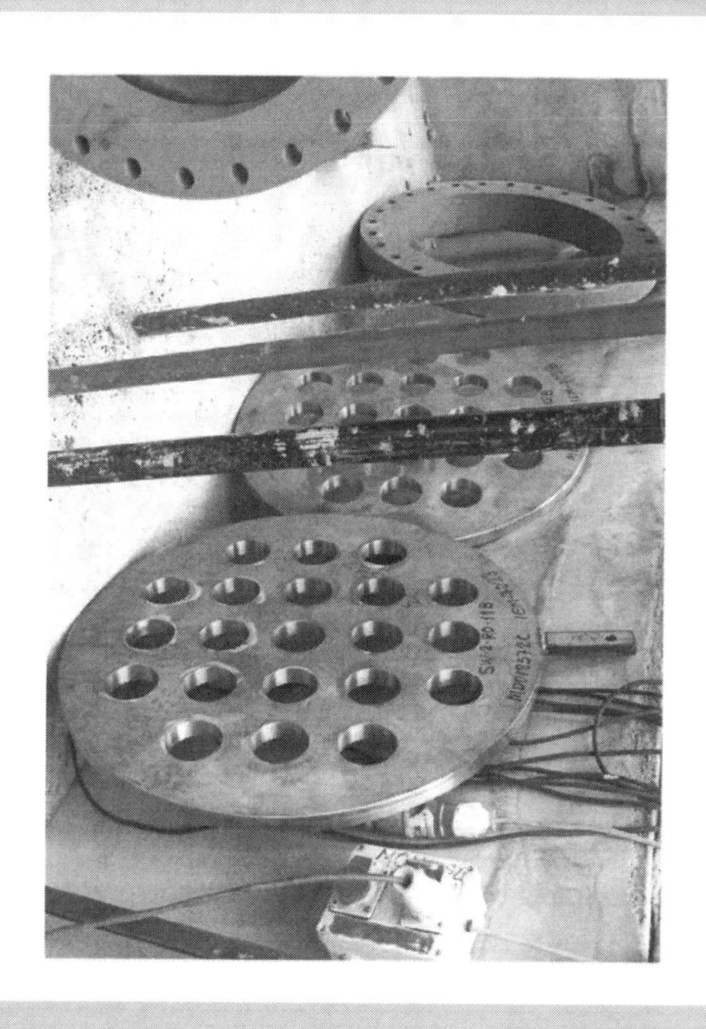

3. DESIGN FOR LIQUIDS

3.1 Relation among the flow rate, the pressure drop and the dimensions of the plate

The flow rate, the pressure drop and the RO dimensions when the fluid is a liquid, are related by the following equations (2) and (3):

$$P_1 - P_o = (P_1 - P_2)/F = 0.64Q^2/v_eC^2d_e^4 \qquad (2)$$

$$Q = 1.25Cd_e^2[v_e(P_1 - P_o)]^{0.5} \qquad (3)$$

P_1 = Pressure at the RO entrance (kg/cm^2 abs.)

P_o = Pressure at the RO holes (kg/cm^2 abs.)

P_2 = Pressure downstream the RO (kg/cm^2 abs.)

F = Pressure recovery factor of the RO plate from the Figure 2

Q = Flow rate (m^3/h)

v_e = Specific volume of the fluid at the RO entrance (m^3/kg)

C = Flow coefficient of the RO from the Figure 1(**Note)**

d_e = Equivalent diameter of the RO from the equation (1) (mm)

The minimum distance between 2 plates in series, to avoid hydraulic interferences, is taken from the Figure 3. This distance is conservative in order to avoid the hydraulic influence of the first plate on the second. It is based on the References [1] and [10].

If it is necessary to install more than 2 plates in series, change to the design of a RO with N stages as is exposed in another Guide.

Note: In the Figure 1 obtain Re from Re = 1273Q/mv$_e$D with the dynamic viscosity m(kg/m-h) and the piping internal diameter D in mm.

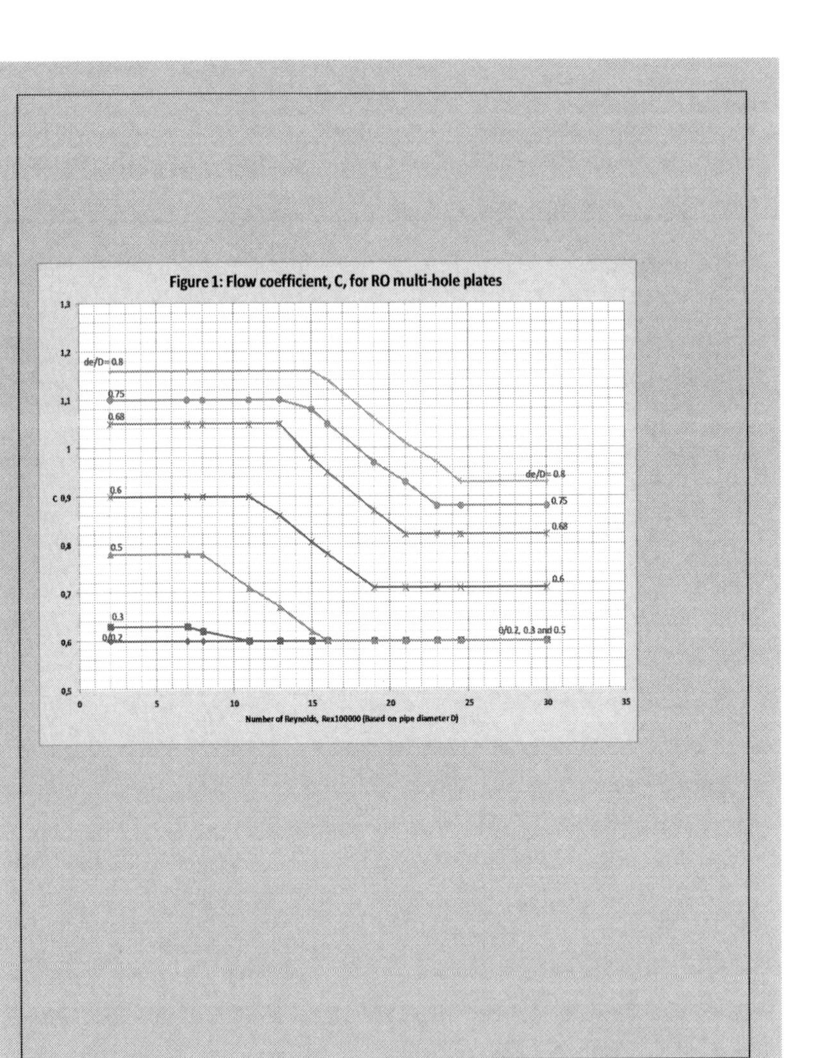

Figure 1: Flow coefficient, C, for RO multi-hole plates

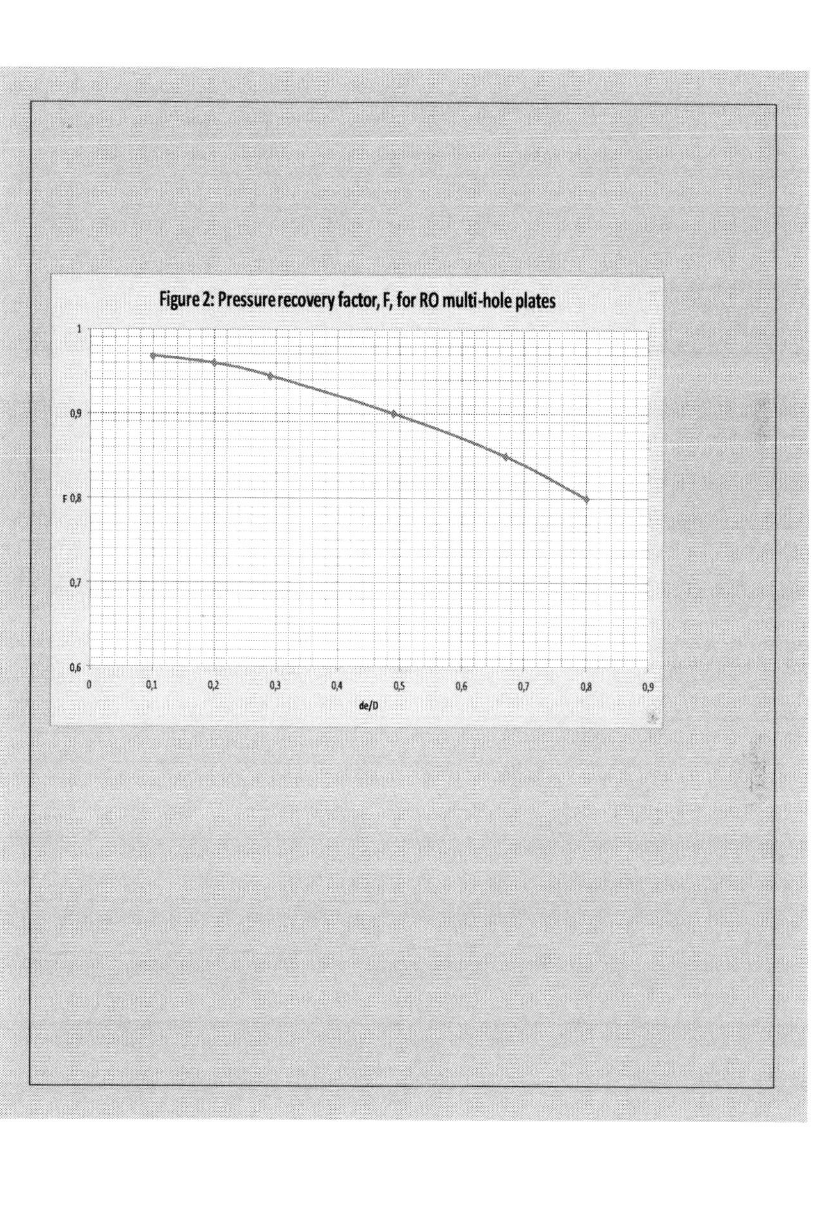

Figure 2: Pressure recovery factor, F, for RO multi-hole plates

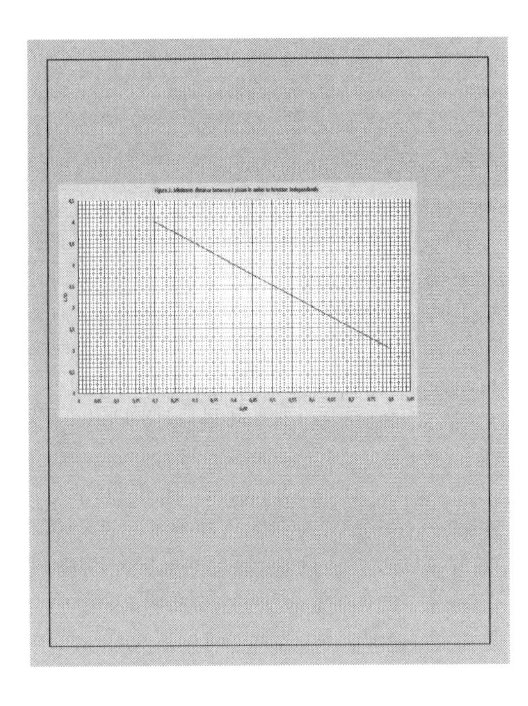

3.2 Cavitation

The incipient cavitation coefficients C_i of the multi-hole plates are showed in the Figure 4. They have been obtained in the tests described in the point 2.

To avoid the cavitation in the RO it is necessary that the cavitation index s be greater than C_i.

The cavitation index s is calculated by the following equation:

$$s = (P_1 - P_v)/(P_1 - P_2) \qquad (4)$$

Therefore, when the RO has been sized according to the point 3.1 it is necessary that it has no cavitation. To know if the RO has cavitation or not, compare C_i with s.

If s is equal or less than C_i the RO has cavitation and to avoid it, divide the total pressure drop to be produced in two stages and design 2 RO in series.

If were necessary 3 or more plates in series to avoid the cavitation, it is recommendable to change the design to a multi-stage RO as shows the Reference [13].

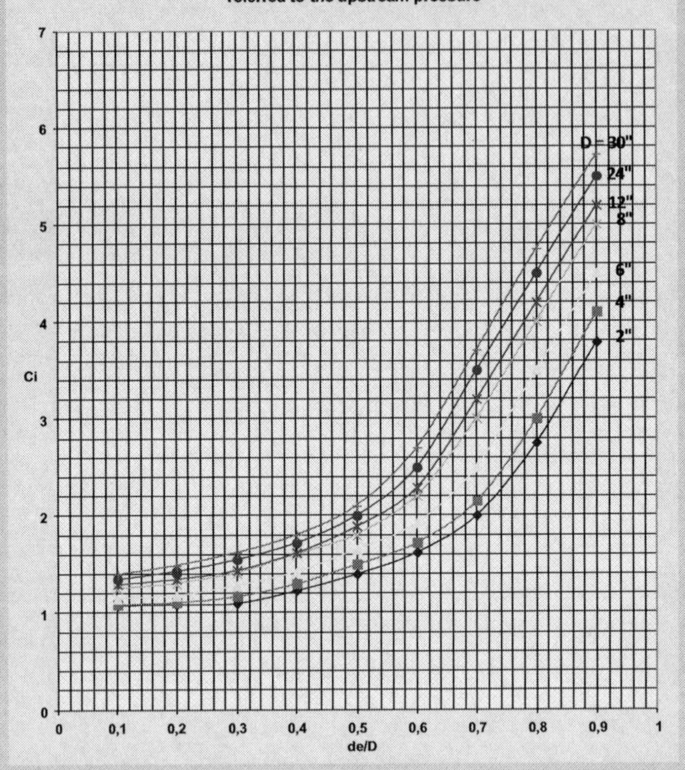

Figure 4: Incipient cavitation coefficient, Ci , for RO multi-hole plates
referred to the upstream pressure
7
6
5
4
3
2
1
0
Ci
0
0,1
0,2
0,3
0,4
0,5
0,6
0,7
0,8
0,9
1
de/D
D = 30"
24"
12"
8"
6"
4"
2"

4. DESIGN FOR SATURATED WATER, STEAM AND GASES

As it is stated in the point 1, the RO multi-hole plates when have sharp-edged holes and the relation between the plate thickness and the equivalent diameter of the holes, t/d_e, is less than 1, may have critical conditions at relative high pressure drops, for compressible fluids passing through them.

4.1 Design for saturated water

The equations that must be applied are also (2) and (3).

Taking into account the Reference [12] and assuming that each hole of the multi-hole plate has a similar behavior as the hole of the 1-hole plates, it´s possible to determine the critical pressure of the saturated water at the holes exit, as show the Figures 5 and 6 for thin plates with n = 13 holes and n = 21 holes.

To calculate the critical flow rate, apply the equation (3) with $P_o = P_{CR}$.

Note that for plates with $d_e/D < 0.50$, the critical pressure is very low compared with the saturation pressure before the plate and therefore, practically there is no critical pressure.

The Figures 5 and 6, give the critical pressure of the saturated water at the exit of the plate holes when the water before the plate is saturated with a vapor content $x_1 = 0$.

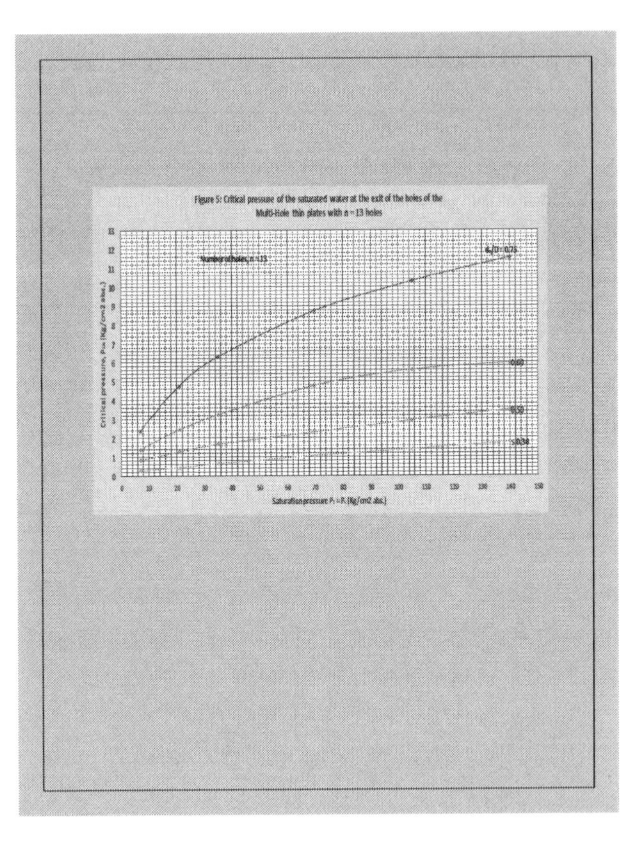
Figure 5: Critical pressure of the saturated water at the exit of the holes of the
Multi-Hole thin plates with n = 13 holes
Number of holes, n = 13
0.60
0.50
0.38
0 10 20 30 40 50 60 70 80 90 100 110 120 130 140 150
0 1 2 3 4 5 6 7 8 9 10 11 12 13

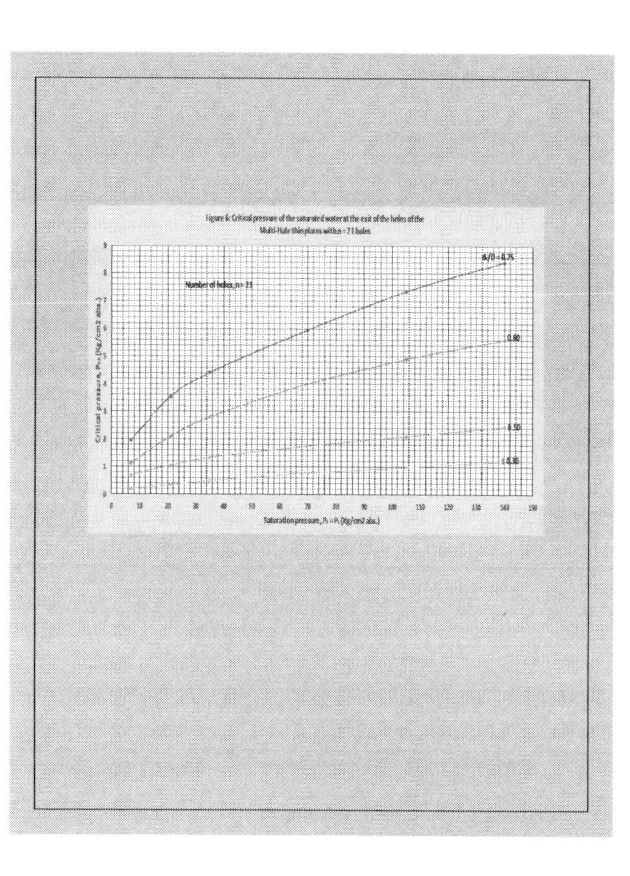
Figure 6: Critical pressure of the saturated water at the exit of the holes of the
Multi-Hole thin plates with n = 21 holes.
Number of holes, n= 21
0.80
0.50
Critical pressure
Saturation pressure
0 10 20 30 40 50 60 70 80 90 100 110 120 130 140 150
0 1 2 3 4 5 6 7 8 9

The velocity of the fluid in the pipe downstream the RO must be limited to a maximum of 25 m/s for continuous operation and 30 m/s for intermittent operation.

The velocity in the pipe may be calculated applying the following equation:

$$\mathbf{V_2 = 3.54 v_{e2} W/D^2} \qquad (5)$$

V_2 = Velocity of the fluid in the pipe downstream the RO (m/s)

W = Flow rate (kg/h)

D = Internal diameter of the pipe (mm)

v_{e2} = Specific volume of the fluid in the downstream pipe (m^3/kg)

To apply this equation, we must know v_{e2}. It can be calculated assuming an isentropic expansion in the RO.

NOTE: The isentropic and isenthalpic expansions of the saturated water, are very similar.

For saturated water, v_{e2} is taken from the steam tables considering the same entropy for $P_1 = P_s$ and $P_2 = P_o$ and the content of steam at 2.

In the Figure 7 it is possible obtain directly the % of steam formed at 2, named in the graph as x.

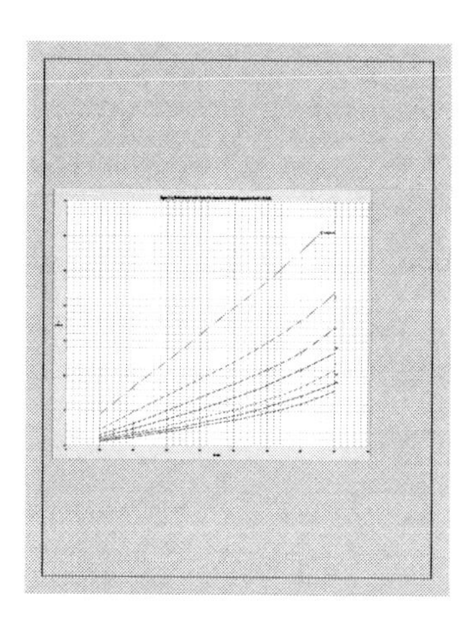

When $P_1 < P_s$, the saturated water before the plate has a vapor content, $x_1 > 0$ and according to [4] and [12] the equations (2) and (3) change to:

$$P_1 - P_o = 0.32Q^2/v_{e1}C^2d_e^4 \qquad (2a)$$

$$Q = 1.77Cd_e^2[v_{e1}(P_1 - P_o)]^{0.5} \qquad (3a)$$

The critical pressure at the exit of the holes, P_{CR}, is that obtained from the Figures 5 and 6, multiplied by the reduction factor R, given in the Figure 8.

Knowing P_1 and x_1, calculate the corresponding value of P_s using the Figure 7, assuming a value of P_s and selecting those that corresponding to P_1 and x_1. Enter with the calculated P_s in the Figures 5 or 6 and obtain P_{CR} and the final critical pressure will be RP_{CR}.

As $R < 1$, the critical pressure of the saturated water with $x_1 > 0$ is still lower than with $x_1 = 0$.

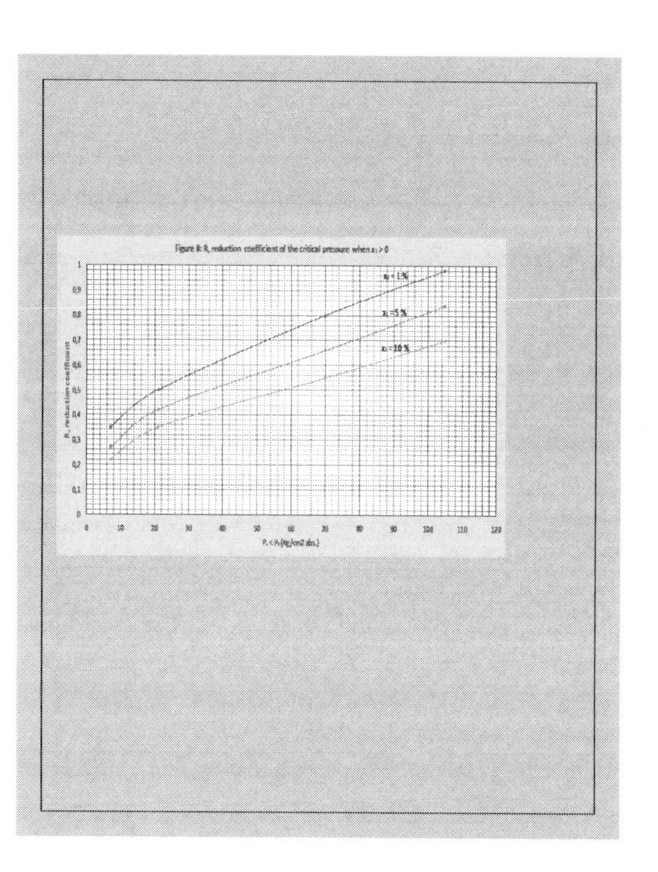

Figure 8: R, reduction coefficient of the critical pressure when $x_1 > 0$

4.2 Design for steam and gases

The equations that relate the pressure drop, the flow rate and the dimensions of the plate are similar to (2) and (3) for liquids adding the expansion factor Y and expressing the volumetric flow Q as mass flow W, being $Q = v_eW$. The equations are:

$$P_1 - P_o = 0.64v_eW^2/Y^2C^2d_e^4 \quad (6)$$

$$W = 1.25Cd_e^2Y[(P_1 - P_o)/v_e]^{0.5} \quad (7)$$

C is obtained from the Figure 1 and Y from the Figures 9, 10 and 11.

The relation of the pressures and specific volumes before and after the RO is given by the following equation:

$$P_1v_{e1}^{\gamma} = P_2v_{e2}^{\gamma} \quad (8)$$

γ is the relation of the specific volumes at constant pressure and constant volume and it is 1.1 for saturated steam and some gases as butane, ethane, hexane, propane etc., 1.3 for reheated steam and other gases as carbon dioxide, ammonia, nitrogen oxide, methane, natural gas etc. and 1.4 for gases as air, hydrogen, nitrogen, carbon monoxide etc.

When it is necessary to install 2 or more plates in series, use the Figure 3 to obtain the minimum straight distance between them.

If it is necessary to install more than 2 plates in series consider changing the design to a multi-stage RO.

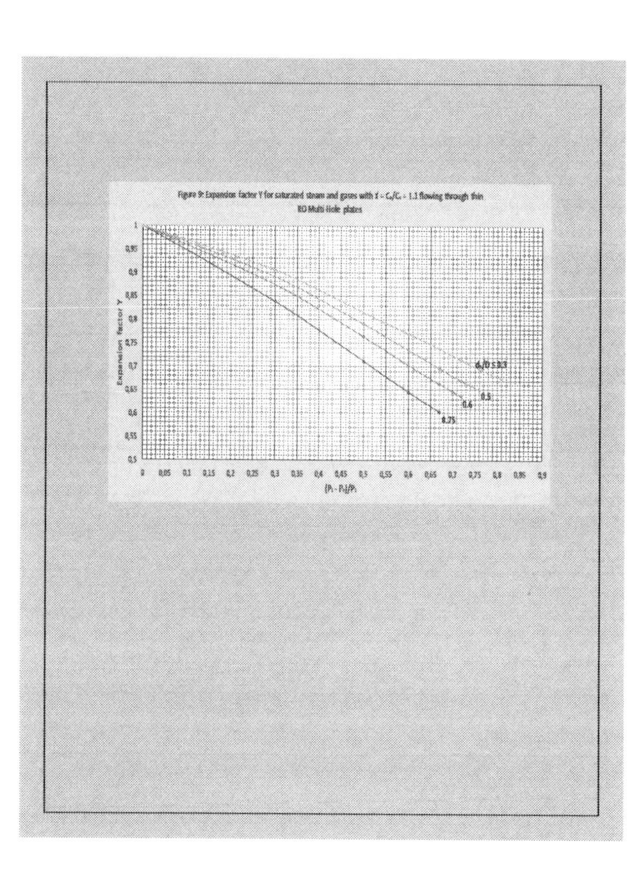

Figure 9: Expansion factor Y for saturated steam and gases with γ = Cp/Cv = 1.3 flowing through thin
RO Multi Hole plates
Expansion factor Y
1
0,95
0,9
0,85
0,8
0,75
0,7
0,65
0,6
0,55
0,5
d/D ≤ 0.3
0.5
0.6
0.75
0
0,05
0,1
0,15
0,2
0,25
0,3
0,35
0,4
0,45
0,5
0,55
0,6
0,65
0,7
0,75
0,8
0,85
0,9
(P1 - P2)/P1

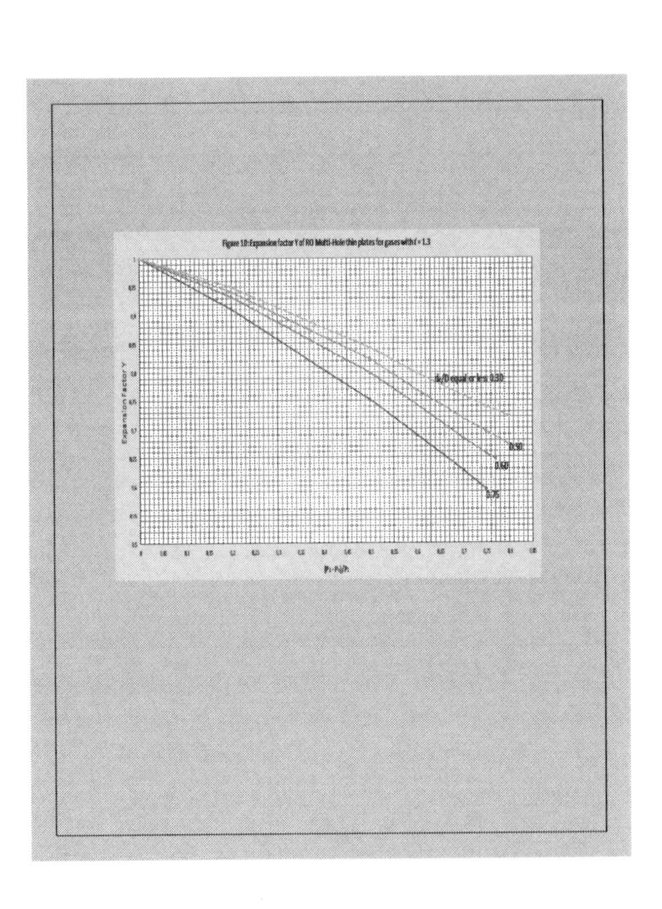

Figure 10: Expansion factor Y of RO Multi-Hole thin plates for gases with k = 1.3

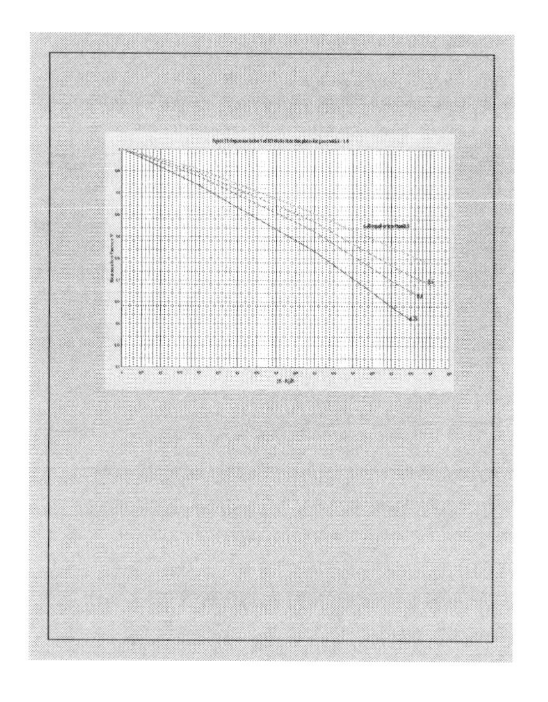

4.3 Noise

The RO must no produce noise and its operation zone in the Figure 9 will be in that named as "without noise fatigue failures".

To confirm this, enter in the Figure 9 with D_2/t_2 in the horizontal axis of the graph and M_2DP in the vertical axis.

D_2 is the internal diameter of the piping downstream the RO and t_2 its thickness, both in the same dimensions.

M_2 is the Mach number of the fluid in the piping downstream the RO and $DP = P_1 - P_o$ the total pressure drop in the RO.

The Mach number is calculated with the following equation:

$$\mathbf{M_2 = 1.13(W/D_2^2)(v_{e2}/P_2\gamma)^{0.5}} \qquad (9)$$

W = Flow rate (kg/h)

D_2 = Internal diameter of the piping downstream the RO (mm)

v_{e2} = Specific volume of the fluid downstream the RO (m^3/kg)

P_2 = Pressure downstream the RO (kg/cm^2 abs.)

γ = Relation of specific heats, C_P/C_V

If the crossing point between the horizontal and vertical entry in the graph is in the zone with pipe failures, it's necessary to reduce the term M_2DP, installing more than 1 RO or decrease the term D_2/t_2 increasing the piping thickness.

Note that a multi-hole plate has a value of P_o greater than a single hole plate for the same $P_1 - P_2$ and W, so the noise provoked by the multi-holes plates is lower than that provoked by the single hole plates. See the Reference [6].

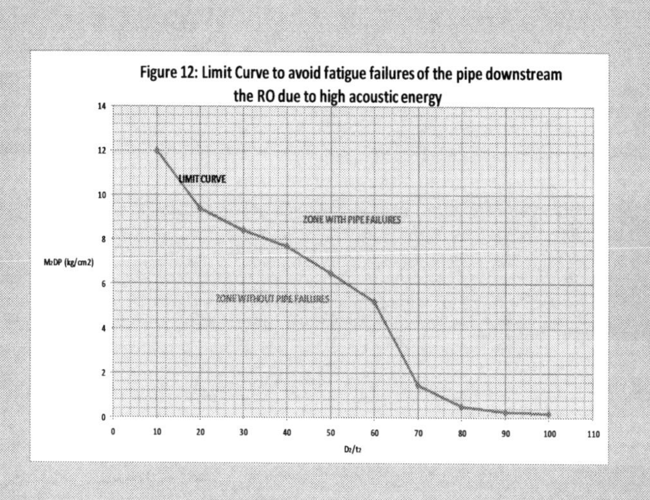

Figure 12: Limit Curve to avoid fatigue failures of the pipe downstream the RO due to high acoustic energy

5. STRUCTURAL DESIGN

The minimum thickness of the plate to avoid its bending may be calculated applying this equation:

$$t = [0.3(P_1 - P_o)D^2R/S_M(R - d_o)]^{0.5} \quad (10)$$

t = Plate thickness (mm)

$P_1 - P_o$ = Overall pressure drop in the RO (kg/cm^2)

D = External plate diameter (mm)

R = Minimum straight distance between the center of two adjacent holes (mm)

S_M = Allowable stress of the plate material (kg/cm^2)

d_o =Diameter of the holes (mm)

The plate material must be stainless steel or alloy steel as SA-240 Tp 304, 304L or 316 with $S_M = 825\ kg/cm^2$

The first natural vibration frequency of the plate, f_1, is calculated with the following equation:

$$f_1 = 1.555x10^6Nt/a^2 \quad (11)$$

N is a coefficient that varies between 0.2 and 2 and f_1 is calculated for N = 0.2 that corresponds to the diametrical fibers of the plate with more holes and N = 2 that corresponds to the diametrical fibers that have only the central hole.

The parameter a is equal to D/2.

The second natural frequency, f_2, is $f_2 = 4.2f_1$

The equation (11) refers to the air. When the fluid is other than air, with the specific volume v_e (m^3/kg) f_1 must be multiplied by this term:

$1/(1 + 0.857x10^{-4}a/v_e t)^{0.5}$

If the natural frequencies f_1 and f_2 are similar to the exciting frequency, change the plate thickness to avoid the coincidence and the resonance of the plate.

When the pressure drop is high, consider the installation of a stiff support or an anchor near the RO to eliminate the piping movements induced by the flow.

6. REFERENCES

[1] E. Casado. "Look at orifice plates to cut piping noise, cavitation". POWER. September 1991

[2] E. Casado. "Avoid vibration, noise and cavitation in pipes using multi-holes plates". INGENIERIA QUIMICA. December 1991.

[3] E. Casado. "Behavior experience of the multi-hole restriction plates". INGENIERIA QUIMICA. October 2003.

[4] .M. W. Benjamin and J. G. Miller. "The Flow of Saturated Water Through Throttling Orifices". Transactions of the ASME. July 1941.

[5] T. J. Rholoff and I. Catton. "Low Pressure Differential Discharge Characteristics of Saturated Liquids Passing Through Orifices". Transactions of the ASME. September 1996.

[6] P. Testud, P. Mousson, A. Hirschberg and Y. Auregane. "Noise generated by cavitating single-hole and multi-hole orifices in water pipe". Journal of Fluids and Structures. 23(2007)163-189.

[7]W. C. Young and R. G. Buydnas. "Roark's Formulas for Stress and Strain". McGraw-Hill seventh Edition.

[8] S. Timoshenko. "Strength of Materials". Espasa- Calpe S. A. 1957.

[9] S. Timoshenko. "Vibration Problems in Engineering". D. Van Nostrand Company Inc. May 1946.

[10] D. María La Rosa, M. María Rossi, G. Ferrarese and S. Malavasi. "On the pressure losses through multi-stage perforated plates". Journal of Fluids Engineering, January 2021.

[11] E. Casado. "Design guide for restriction orifice plates with 1 central hole". Revision 2. Amazon, July 2021.

[12] E. Casado. "The flow of saturated water through restriction orifice thin plates". Revision 3. Research Gate, June 2021.

[13] E. Casado. "Design guide for Multi-Stage restriction orifice (ROs) plates with 1 hole and n > 3 holes". Amazon, July 2021.

CHAPTER 3

DESIGN GUIDE FOR MULTI-STAGE RO PLATES WITH 1 SINGLE HOLE AND MORE THAN 3 HOLES

DESIGN GUIDE FOR MULTI-STAGE RO PLATES WITH 1 SINGLE HOLE AND MORE THAN 3 HOLES

(Fluids: liquids, saturated water, steam and gases)

July 2021. Revision 2

Emilio Casado Flores
Power Consulting Engineer

CONTENT

1. INTRODUCTION

These types of ROs has application when it is necessary divide the pressure drop in several steps in order to avoid the cavitation and noise if only one or two plates were installed and the cavitation persists and the noise is high.

See, for example, the Figure 1 that represents a multi-stage RO with plates that have one hole and the Figure 2 that represents a multi-stage RO with plates of several holes.

The multi-stage ROs may be installed for example, in the minimum flow recirculation lines of the high-pressure pumps or in steam and gas pipes, where it is necessary a high reduction of the pressure.

Usually, for liquids, the last stage N has the hole or holes with a greater diameter than the other stages. For steam and gases, if each stage is designed for the same pressure drop, the diameter of the holes increases in each stage.

The holes are sharp-edged and the plates are thin plates, that is the relation between its thickness and the diameter of the hole t/d_o for the one-hole plates or t/d_e for the n holes plates, is less than 1.

The known data to design the RO are the RO upstream pressure, P_1, the total pressure drop in the RO, $P_1 - P_2$ and the flow rate Q. With these data we must calculate the number of stages N, the hole diameters d_o and D_o for the plates that have only 1 hole, the hole equivalent diameters d_e and D_e if each plate has $n > 3$ holes and the RO total length L_T.

See the Reference [6].

2. DESIGN FOR LIQUIDS

It is necessary to design each stage without cavitation. To do this, the design begins calculating the last plate N.

If the RO has the plates of one hole, assume that the hole of the last plate has a diameter of D_o and that we name P_N as the pressure before this last plate. We establish that the cavitation index of this last plate is a little greater than its incipient cavitation coefficient, C_i, for example 10 %, so it is possible to write these equations:

$$(P_N - P_V)/(P_N - P_2) = 1.1C_i \qquad (1)$$

$$P_N = (1.1C_iP_2 - P_V)/(1.1C_i - 1) \qquad (2)$$

P_N = Pressure before the last plate (kg/cm^2 abs.)

P_V = Vapor pressure of the fluid (kg/cm^2 abs.)

P_2 = Downstream RO pressure (kg/cm^2 abs.)

C_i = Incipient cavitation coefficient

C_i is obtained in the Figure 5 for plates with n = 1 hole and in the Figure 8 for plates with n > 3 holes (multi-hole plates) and they are valid for whichever be P_N that is the pressure before the plate. See Reference [2].

Note: The cavitation index s is defined as $s = (P_N - P_V)/(P_N - P_2)$ in the equation (1).

If the design of the RO is with plates of one hole, calculate the relation between the pressure drop and the flow rate in the last plate N with the following equation. See the Reference [1].

$$P_N - P_2 = 0.64Q^2F/v_eC^2D_o^4 \qquad (3)$$

Q = Flow rate (m^3/h)

F = Pressure recovery factor of the plate

v_e = Specific volume of the fluid before the plate (m^3/kg)

C = Flow coefficient of the plate

D_o = Hole diameter of the last plate N (mm)

Obtain F in the Figure 3 and C in the Figure 4.

If the RO is designed with plates of several holes, n>3, usually as a rule of thumb consider n = 4, 5 or 9 for D equal or less than 4 in and n = 9, 13 or 21 for D > 4 in. D is the piping diameter.

The relation between the pressure drop and the flow rate is calculated also with the equation (3) but change D_o by D_e that is the equivalent diameter of the holes of the plate. If the plate has n holes and each one has the diameter D_o, D_e is:

$$\mathbf{D_e = D_o n^{0.5}} \qquad (4)$$

See the References [3], [4] and [5].

Obtain F and C for the multi-hole plates in the Figures 6 and 7.

The parameters F and C of the 1 Hole and Multi-Hole plates, correspond to the plate of each stage, assuming that the upstream plate doesn´t have hydraulic influence. To achieve this, the plates must be separated unless the distance L/D = 1, according to the References [17] and [18].

It´s convenient that the plates of 1 Hole, have a bevel of 45º, or at least 30º, at the exit of the hole, as show the sketch of the Figure 1a, in order to ease that the downstream plate be hydraulic independent of the upstream.

See the Figures 1, 1a and 2.

The bevel, as show the Figure 1a, let the outlet jet stream of the hole exit, to rise the internal wall of the pipe, before reach the downstream plate.

For bevels of 45 and 30 degrees, the value of $x/D < 1$.

For example, as shows the sketch of the Figure 1a, is:

$R \times \sin 45 = y = 0.5(D - d_o)$; $R = 0{,}707(D - d_o)$

$R x \cos 45 = x = 0{,}707(D - d_o)0{,}707 = 0{,}5(D - d_o)$

$x/D = 0{,}5(1 - d_o/D) < 1$

Therefore $L/D = 1 > x/D$ and the separation distance of $L/D = 1$, is enough, because the flow stream reaches the internal wall of the pipe before arrive to the next plate.

For a bevel of 30 degrees, even x is a little greater than before, also, $x/D < 1$.

For the Multi-Hole plates, the Reference [18], shows that $L/D = 1$, is enough without do the bevel, but including a bevel, could be recommendable.

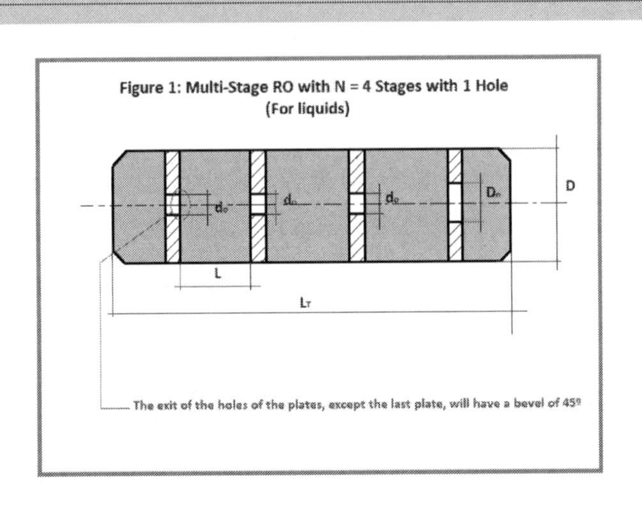

Figure 1: Multi-Stage RO with N = 4 Stages with 1 Hole
(For liquids)

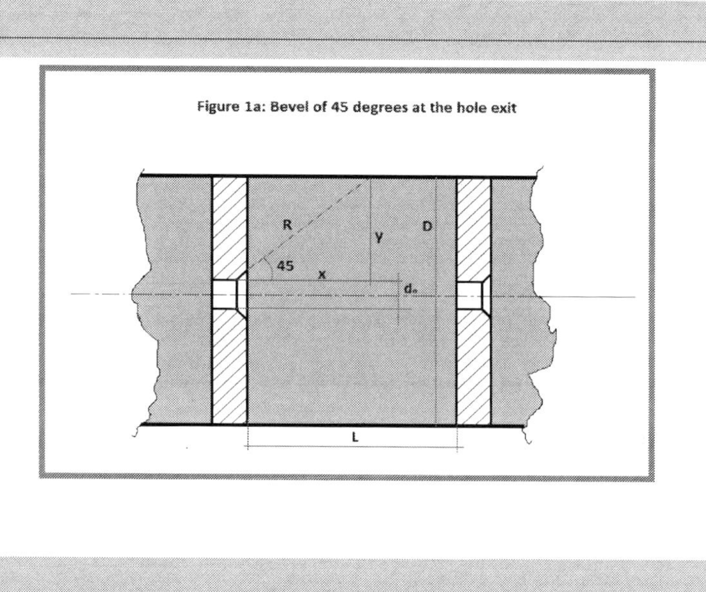

Figure 1a: Bevel of 45 degrees at the hole exit

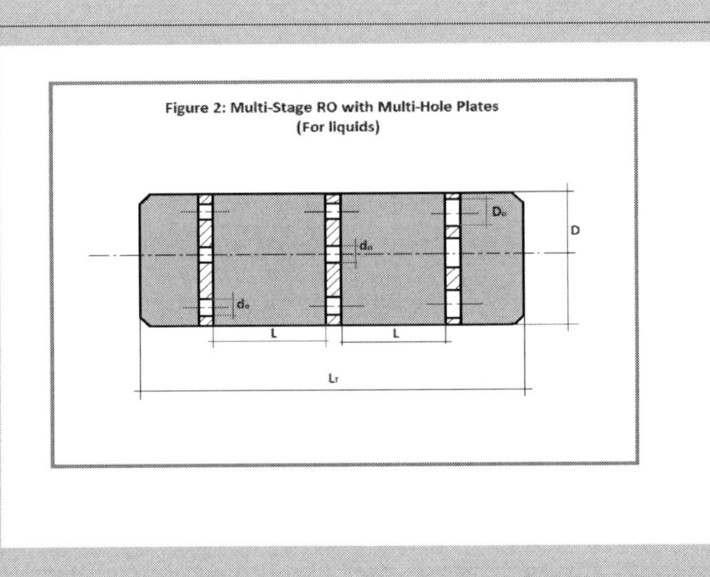

Figure 2: Multi-Stage RO with Multi-Hole Plates
(For liquids)

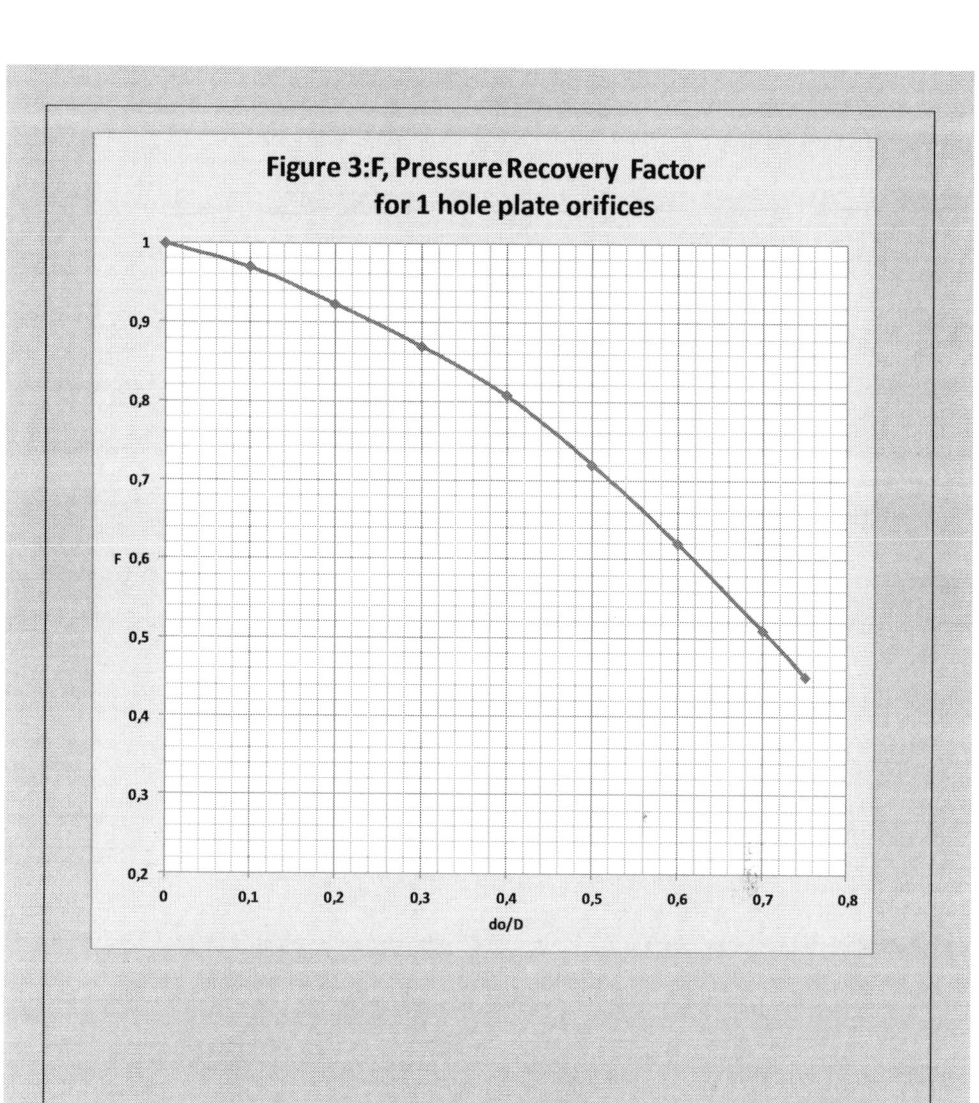

Figure 3:F, Pressure Recovery Factor
for 1 hole plate orifices
1
0,9
0,8
0,7
F 0,6
0,5
0,4
0,3
0,2
0
0,1
0,2
0,3
0,4
0,5
0,6
0,7
0,8
do/D

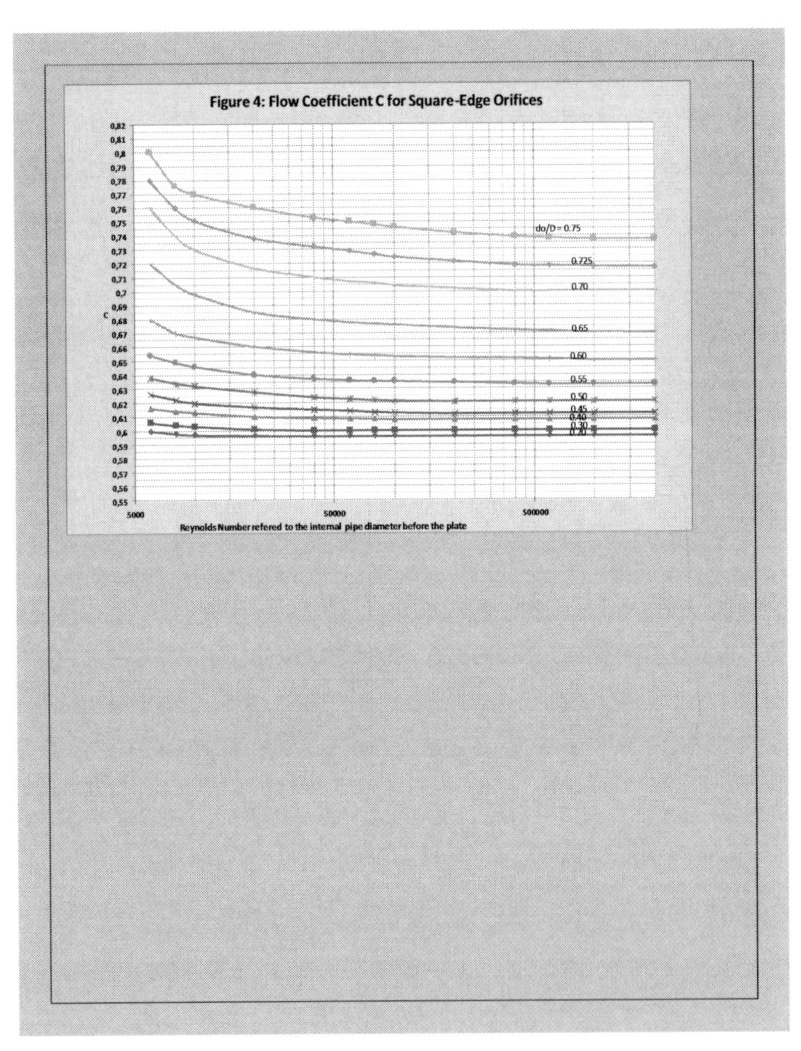

Figure 4: Flow Coefficient C for Square-Edge Orifices
0,82
0,81
0,8
0,79
0,78
0,77
0,76
0,75
0,74
0,73
0,72
0,71
0,7
0,69
0,68
0,67
0,66
0,65
0,64
0,63
0,62
0,61
0,6
0,59
0,58
0,57
0,56
0,55
C
do/D = 0.75
0.725
0.70
0.65
0.60
0.55
0.50
0.45
0.40
0.30
0.20
5000
50000
500000
Reynolds Number refered to the internal pipe diameter before the plate

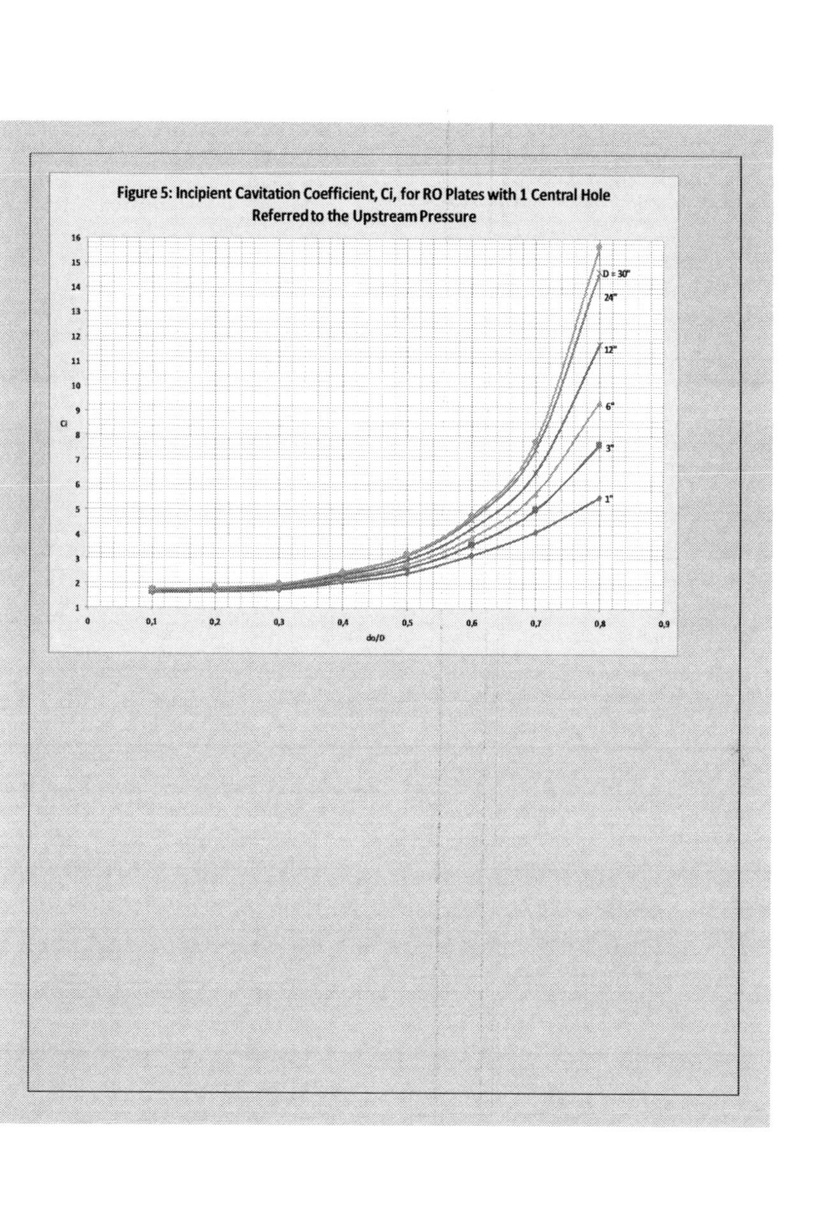

Figure 5: Incipient Cavitation Coefficient, Ci, for RO Plates with 1 Central Hole
Referred to the Upstream Pressure
Ci
do/D
D = 30"
24"
12"
6"
3"
1"
0
0,1
0,2
0,3
0,4
0,5
0,6
0,7
0,8
0,9
1
2
3
4
5
6
7
8
9
10
11
12
13
14
15
16

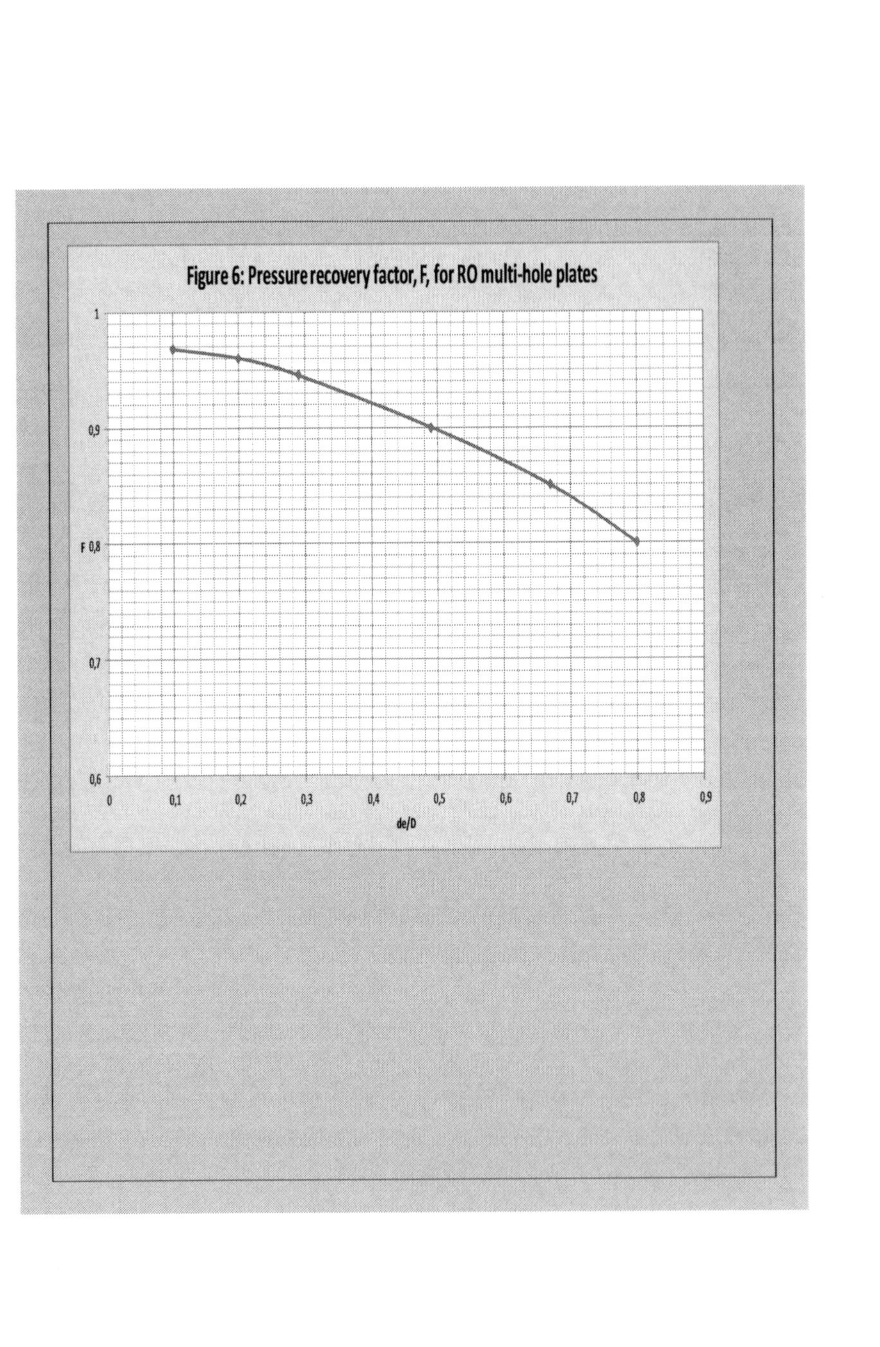
Figure 6: Pressure recovery factor, F, for RO multi-hole plates
1
0,9
F 0,8
0,7
0,6
0
0,1
0,2
0,3
0,4
0,5
0,6
0,7
0,8
0,9
de/D

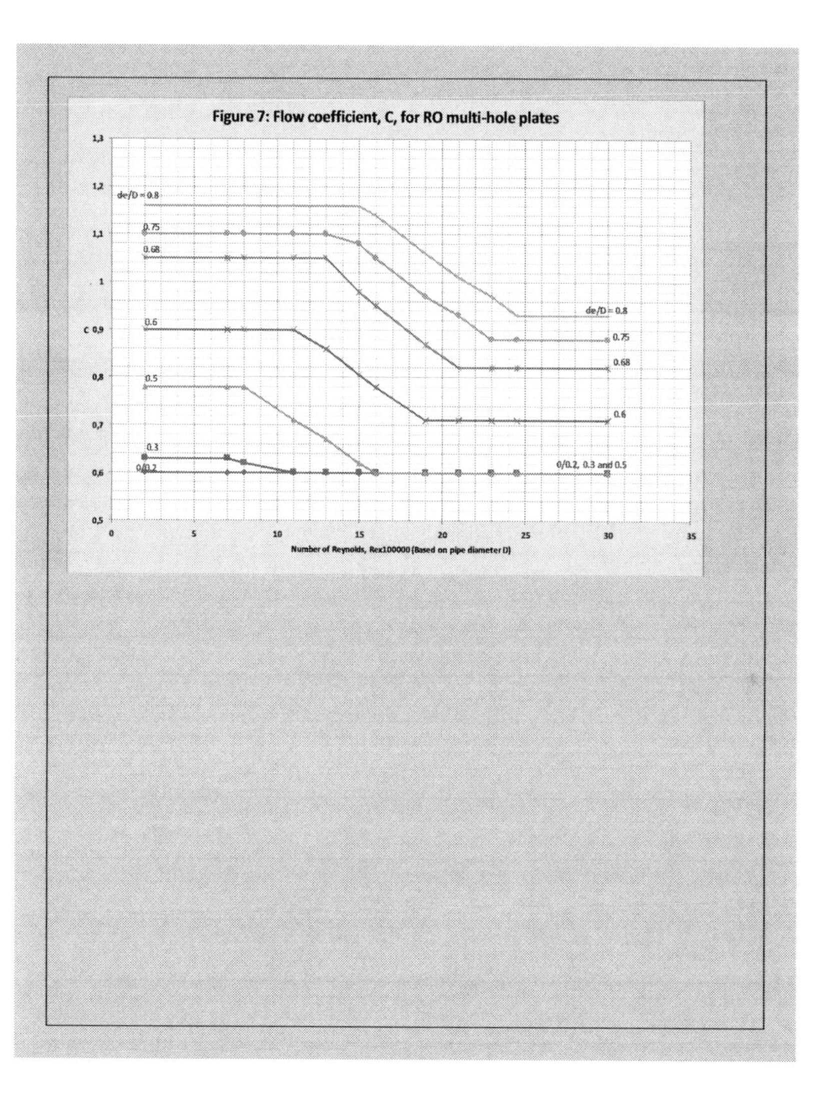

Figure 7: Flow coefficient, C, for RO multi-hole plates

Figure 8: Incipient cavitation coefficient, Ci , for RO multi-hole plates referred to the upstream pressure

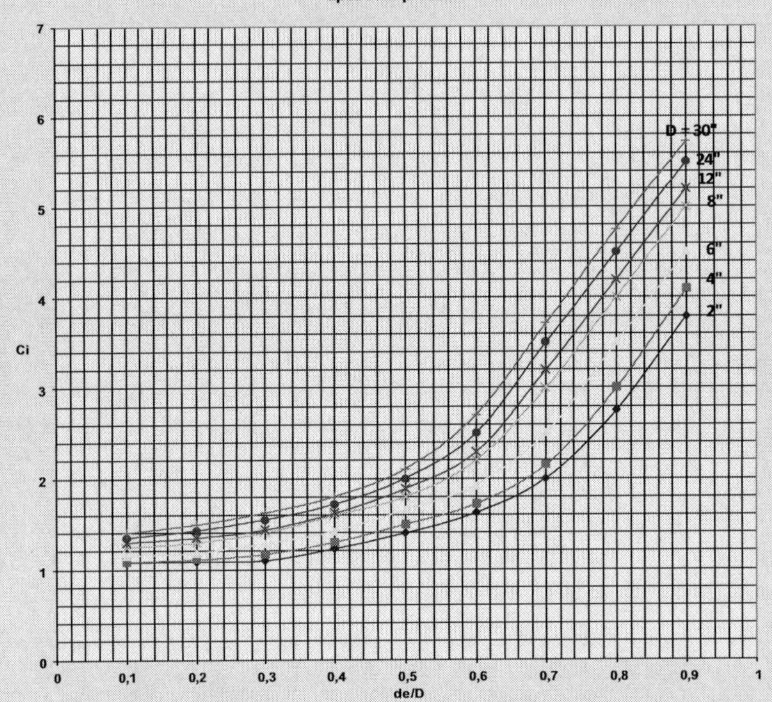

With the equations (2) and (3) we obtain for the ROs designed with plates of one hole:

$$(P_2 - P_V)/(1.1C_i - 1) = 0.64Q^2F/v_eC^2D_o^4 \qquad (5)$$

For the ROs designed with plates of n holes, the equation will be:

$$(P_2 - P_V)/(1.1C_i - 1) = 0.64Q^2F/v_eC^2D_e^4 \qquad (6)$$

Resolve these equations by trials. The known data are P_2, P_V, the internal piping diameter D and Q. Give a value to D_o or D_e (previously we have selected n) and obtain the corresponding values of F, C and C_i in the Figures 3, 4 and 5 or 6, 7 and 8 respectively.

Check if these values fulfill the equations (5) or (6) if not try other value of D_o or D_e and repeat the process.

Now obtain P_N applying directly the equation (2).

The rest of the stages, will be plates with the same hole diameter, d_o for one-hole plates and $d_e = d_o n^{0.5}$ for n holes plates.

As the rest of the plates have the same dimensions, they will have also the same pressure drop that is $(P_1 - P_N)/(N - 1)$.

Note that still we do not know N.

The plate N – 1 is designed as the plate N imposing that $s = 1.1C_i$

As $P_N > P_2$, will be $d_o < D$.

Therefore, it will be as in the equation (3):

$$(P_N - P_V)/(1.1C_i - 1) = 0.64Q^2F/v_eC^2d_o^4 \qquad (7)$$

$$(P_N - P_V)/(1.1C_i - 1) = 0.64Q^2F/v_eC^2d_e^4 \qquad (8)$$

As before, resolve in the same form by trials of d_o or d_e these equations.

Each stage will have the same pressure drop, that is $(P_1 - P_N)/(N - 1)$, so taking into account the equations (7) or (8), the number of stages N to design the RO is given by the following equations for 1-hole plates or n holes plates:

$$\mathbf{N = 1 + (P_1 - P_N)/(0.64Q^2F/v_eC^2d_o^4)} \quad (9)$$

$$\mathbf{N = 1 + (P_1 - P_N)/(0.64Q^2F/v_eC^2d_e^4)} \quad (10)$$

Usually, N is a decimal number. Take for N the next whole number.

For the same pressure drop and flow rate, N is 2 or 3 times lower in the RO with n holes plates than in the RO with one-hole plates. The reason is the lower cavitation coefficients of the n holes plates.

We can define a global incipient cavitation coefficient named C_{iM} for the RO with N stages.

When the RO has the stage N with the hole diameter greater than the other N – 1 stages, as it is exposed before, C_{iM} may be calculated with the following equation:

$$\mathbf{C_{iM} = (C_i + N - 2)C_{iN}/[(N - 1)C_{iN} + C_i - 1]} \quad (11)$$

C_{iN} is the incipient cavitation coefficient of the last stage N and C_i the incipient cavitation coefficient of the other N – 1 stages.

If all the N stages have the same dimensions and number of holes, all them have the same incipient cavitation coefficient C_i. In this case, calculate C_{iM} with the following equation:

$$\mathbf{C_{iM} = (C_i + N - 1)/N} \quad (12)$$

With this method of design, the last plate has $D_o > d_o$ and $D_e > d_e$ and the cavitation is avoided in each stage.

3. DESIGN FOR SATURATED WATER, STEAM AND GASES

As in the point 1, these ROs have thin plates with sharp-edged holes.

The References [15] and [16] show that this type of individual plates may have critical conditions for high values of the pressure drop, around 80 or 90 % of the total pressure drop. As each stage is designed for a pressure drop around 25 or 30 % of the total pressure drop, we consider that there are not critical conditions when the flow through the Multi-Stage RO corresponds to a compressible fluid as saturated water that flashes, steam or gases. See also References [7], [8] and [9].

The separation between the plates may be also 1D, but as the pressure recovery is not relevant, the bevel at the hole exit may be deleted.

3.1 Design for saturated water

The design is the same for ROs with plates of n = 1 hole than for plates with n > 3 holes.

Calculate the plate of each stage to produce the same pressure drop, so the holes size increases with the number of stages. The lowest belongs to the first stage, N = 1 and the biggest to the last stage N. See the Figure 11 for a multi-stage RO with N = 4 stages with plates of 1 hole and the Figure 11a for multi-hole plates.

The pressure drop in each stage will be: $(P_1 - P_F)/N$.

The design of the RO begins with the calculation of the hole diameter d_1 of the first plate applying the equation (3) in this form:

$$\mathbf{P_1 - P_2 = (P_1 - P_F)/N = 0.64 v_{e1} W^2/C_1^2 d_1^4}$$

NOTE: W is the flow rate (Kg/h) and P_F (Kg/cm^2 abs.) the pressure at the RO exit.

As there is no pressure recovery, P_2 is practically the pressure in the hole of this first stage.

P_1 is the saturation pressure at the entrance of the RO.

For n = 1 hole, C_1 is obtained in the Figure 4 and for n > 3 holes, C_1 is obtained in the Figure 7.

For plates with n > 3 holes, change d_1 by d_{e1}.

The before cited equation must be resolved for C_1 and d_1 by trials of d_1. Assume a value for d_1 and obtain the corresponding value of C_1 in the Figure 4. Check if they fulfill the equation. If not, try another value for d_1 until find the pair of values of C_1 and d_1 that fulfill the equation.

When d_1 has been calculated, the diameter of the holes of the other stages, d_2, d_3 and d_4 for the RO of the Figure 9 are calculated with these equations:

$$\mathbf{d_2 = (C_1/C_2)^{0.5}(v_{e2}/v_{e1})^{0.25}d_1x0,84} \quad (13)$$

$$\mathbf{d_3 = (C_1/C_3)^{0.5}(v_{e3}/v_{e1})^{0.25}d_1x0,84} \quad (14)$$

$$\mathbf{d_4 = (C_1/C_4)^{0.5}(v_{e4}/v_{e1})^{0.25}d_1x0,84} \quad (15)$$

These equations come from these other equations (13a), (14a) and (15a):

$$\mathbf{(P_1 - P_F)/N = P_2 - P_3 = 0.32v_{e2}W^2/C_2d_2^4 = P_1 - P_2} \quad (13a)$$

$$\mathbf{(P_1 - P_F)/N = P_3 - P_4 = 0.32v_{e3}W^2/C_3d_3^4 = P_1 - P_2} \quad (14a)$$

$$\mathbf{(P_1 - P_F)/N = P_4 - P_F = 0.32v_{e4}W^2/C_4d_4^4 = P_1 - P_2} \quad (15a)$$

NOTE: According to the References [7], [15] and [16], the pressure drop of each stage, after the first one, is approximately half of this because the saturated water will have vapor content with $x > 0$.

To resolve these equations and calculate d_2, d_3 and d_4 it is necessary to know v_{e2}, v_{e3} and v_{e4}. To calculate the specific volumes before each plate, first we obtain x that it is the % of steam produced in the isentropic expansion from $P_1 = P_S$ in each plate. So, in the first stage, x_1 is the % of steam that has been produced in the plate in the expansion from $P_1 = P_S$ to P_2. In the second stage, x_2 is the steam produced in the first plate more the steam produced in the second plate in the expansion from P_2 to P_3. This is equivalent to the direct expansion from P_1 to P_3.

In the same form, x_3 is the % of steam after the stage N = 3 and it is calculated in the expansion from P_1 to P_4 and x_4 the % of steam after the last stage N = 4.

In the Figure 10 for multi-hole plates will be $d_{ei} = 3d_i$ (n = 9 holes) for each stage i. The equations are similar, changing d_i by d_{ei}.

See the Figures 9 and 10 that show the design of this type of RO for saturated water with the same pressure drop in each stage.

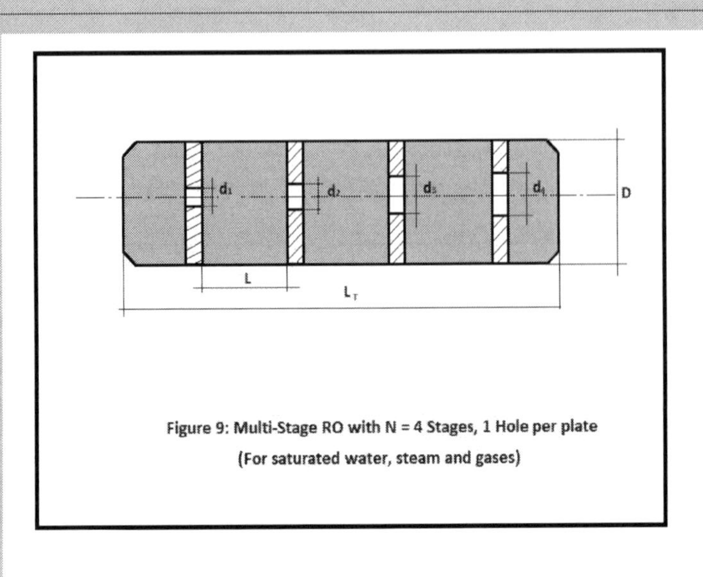

Figure 9: Multi-Stage RO with N = 4 Stages, 1 Hole per plate

(For saturated water, steam and gases)

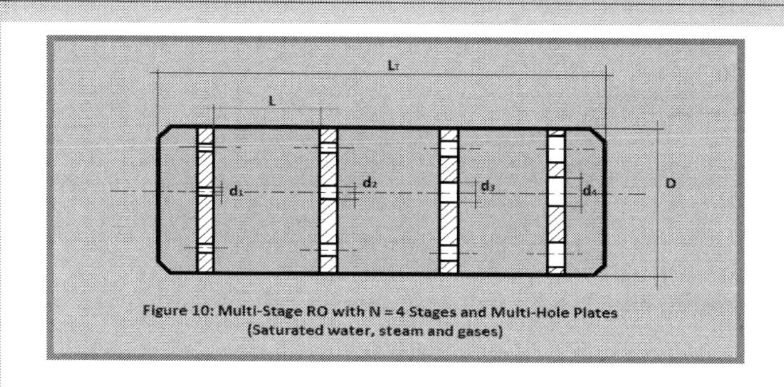

Figure 10: Multi-Stage RO with N = 4 Stages and Multi-Hole Plates
(Saturated water, steam and gases)

The values of x are obtained from the Figure 11 for P_S in the horizontal axis and the parameter $(P_S - P_i)/P_S$ with i = 1, 2, 3, 4, ... N.

The graph of the Figure 11 is valid until P_S = 1000 psia (70.3 kg/cm² abs.).

Alternatively for pressures greater than 1000 psia, use the following equation to calculate x:

$$x = AP_S^B$$

A is obtained from the Figure 12 and $B = -0.2(P_S - P_i)/P_S + 0.54$.

The equation of x is valid also for $P_S > 250$ psia (17.57 kg/cm² abs.)

With the values of x after each stage, the specific volume v_{ei} is calculated using the steam tables and the equation:

$$\mathbf{v_{ei} = v_{ef} + v_{efg}x_i} \quad (16)$$

v_{ei} = Specific volume before the stage i (m³/kg)

v_{ef} = Specific volume of the saturated condensate at P_i (m³/kg)

v_{efg} = Specific volume of the change from condensate to steam (m³/kg)

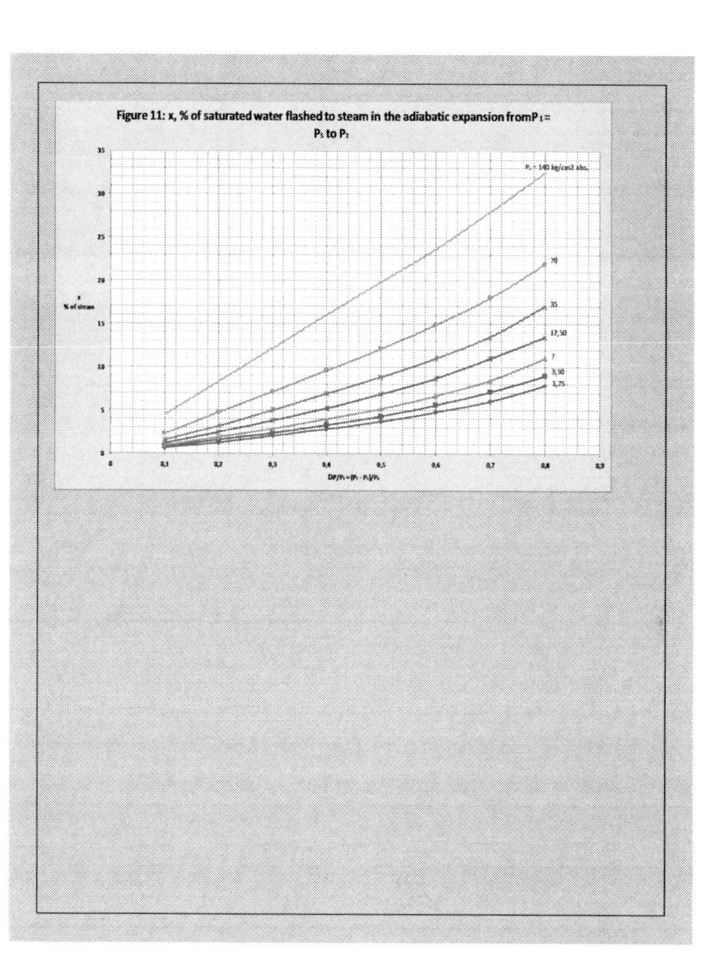
Figure 11: x, % of saturated water flashed to steam in the adiabatic expansion from P1 = Ps to P2
35
30
25
20
15
10
5
0
x
% of steam
0
0,1
0,2
0,3
0,4
0,5
0,6
0,7
0,8
0,9
DP/Ps = (Ps - P2)/Ps
Ps = 140 kg/cm2 abs.
70
35
17,50
7
3,50
1,75

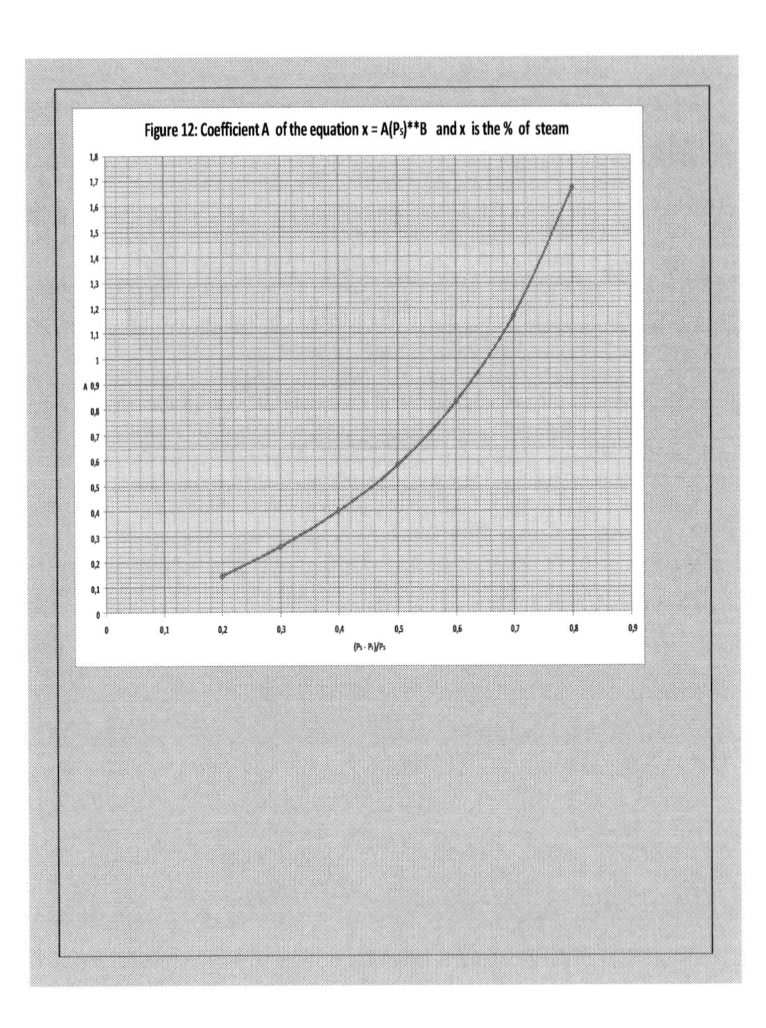

Figure 12: Coefficient A of the equation $x = A(P_s)^{**}B$ and x is the % of steam

Finally resolve the equations (13), (14) and (15) to obtain d_2, d_3 and d_4 for ROs with plates of 1 hole and d_{e2}, d_{e3}, and d_{e4} for ROs with plates that have $n > 3$ holes.

3.2 Design for steam and gases

The relation between the flow rate and the pressure drop is calculated also with the equation (3) including the expansion factor Y. This is:

$$(P_1 - P_F)/N = 0.64v_{e1}W^2/Y^2C^2d_o^4 \qquad (17)$$

The values of Y for saturated steam, reheated steam and gases are obtained from the Figures 13, 14 and 15 for plates with one hole and 13a, 14a and 15a for plates with $n > 3$ holes.

Design each stage to produce the same pressure drop, that is:

$(P_1 - P_F)/N = 0.64v_{e1}W^2/Y_1^2C_1^2d_1^4 = 0.64v_{ei}W^2/Y_i^2C_i^2d_i^4$

The relation between the pressure and the specific volume of each stage is given by the following equation:

$$P_1v_{e1}^{\gamma} = P_2v_{e2}^{\gamma} = \ldots\ldots\ldots\ldots = P_Nv_{eN}^{\gamma} \qquad (18)$$

γ is the relation between the specific heats of the fluid. For saturated steam and some gases as benzene, butane, ethane, pentane, hexane, propane and propylene, is 1.1; for reheated steam and some gases as carbon dioxide, sulfur dioxide, hydrogen sulfide, ammonia, nitrogen oxide, chlorine, methane, acetylene, ethylene and natural gas, is 1.3; for other gases as air, hydrogen, oxygen, nitrogen, carbon monoxide, nitric oxide and hydrochloric acid, is 1.4.

Taking into account the equations (17) and (18) and naming $d_1 = d_o$ the hole diameter of whichever stage is:

$$d_i = (Y_1C_1/Y_iC_i)^{0.5}(P_1/P_i)^{0.25/\gamma}d_o \qquad (19)$$

The diameter d_o of the first stage is calculated directly from the equation (17) in this form:

$d_o = 0.89(W/Y_1C_1)^{0.5}(v_{e1}N/(P_1 - P_F))^{0.25}$

For plates with n > 3 holes change d_o by d_e

Note that for the design of the ROs of N stages for steam and gases, only is necessary to know the specific volume at the entrance of the RO but not between every two stages as in the case of the saturated water.

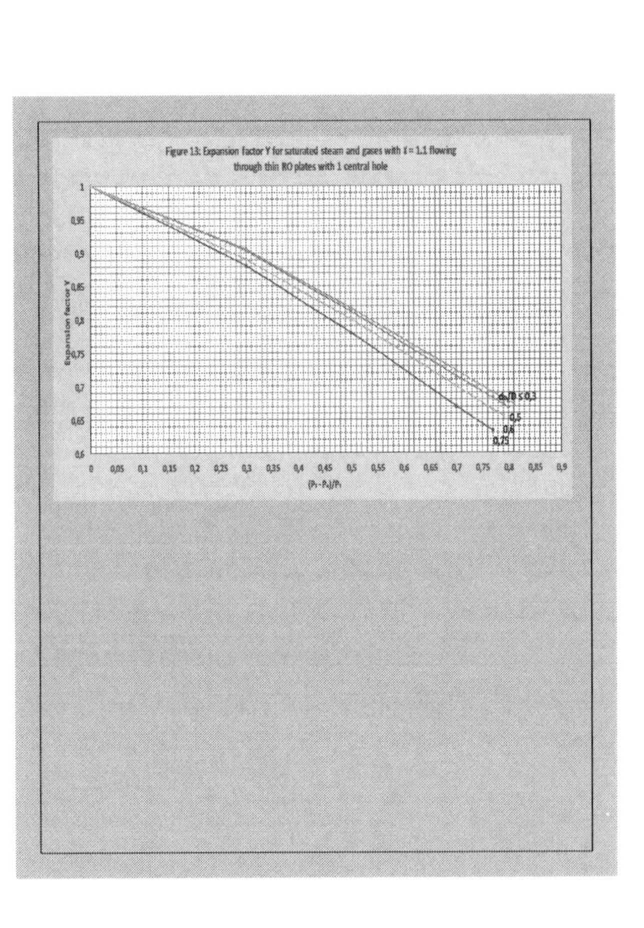

Figure 13: Expansion factor Y for saturated steam and gases with $\kappa = 1.1$ flowing through thin RO plates with 1 central hole

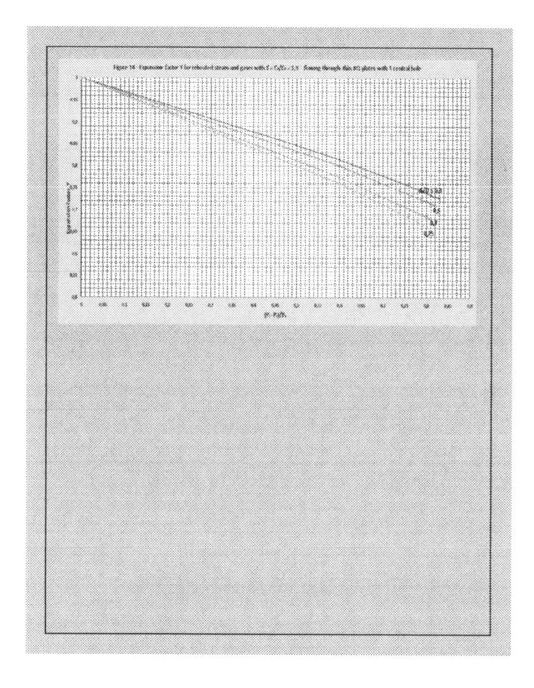

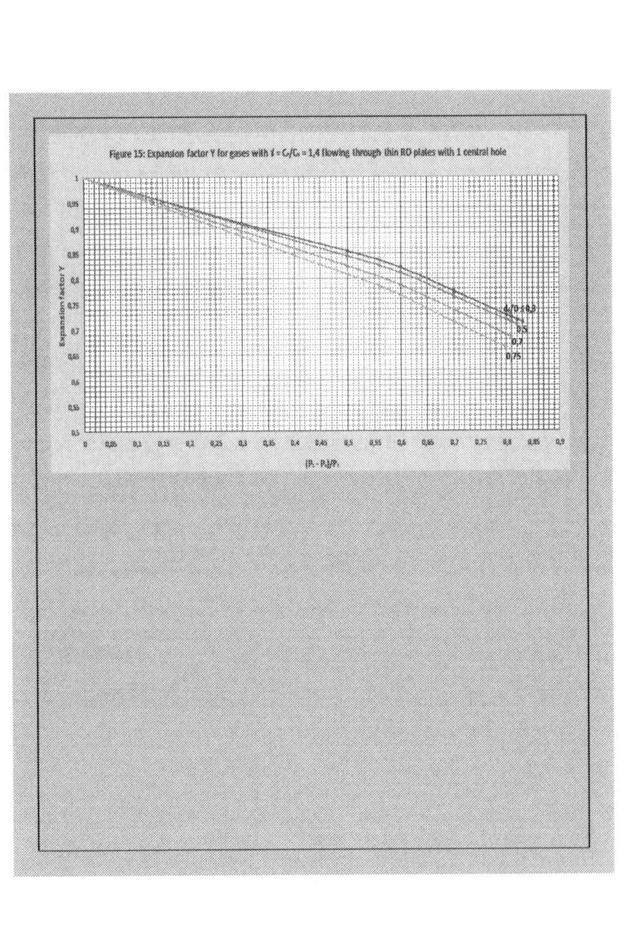

Figure 15: Expansion factor Y for gases with $\gamma = C_p/C_v = 1,4$ flowing through thin RO plates with 1 central hole

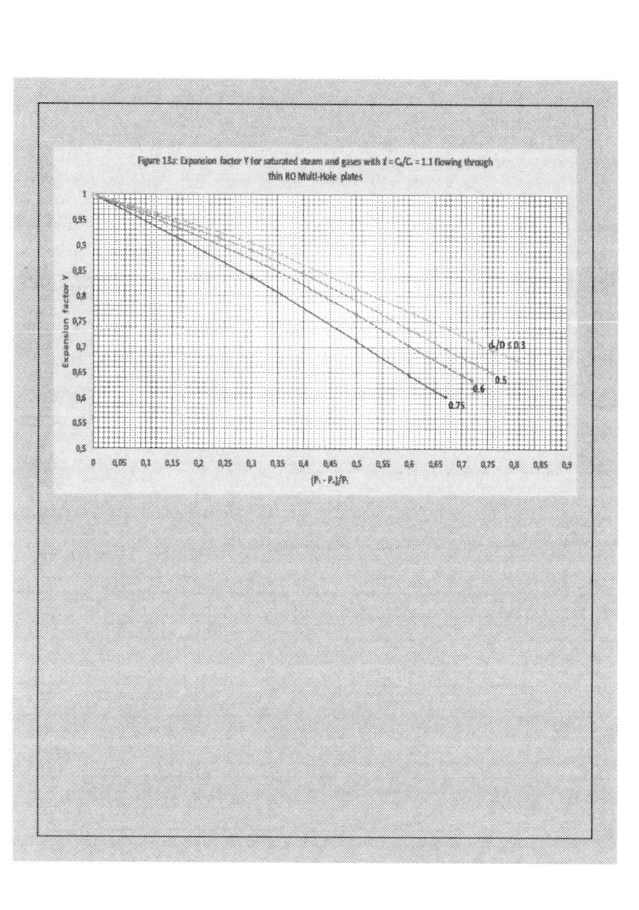

Figure 13a: Expansion factor Y for saturated steam and gases with k = Cp/Cv = 1.1 flowing through
thin RO Multi-Hole plates
Expansion factor Y
1
0,95
0,9
0,85
0,8
0,75
0,7
0,65
0,6
0,55
0,5
0
0,05
0,1
0,15
0,2
0,25
0,3
0,35
0,4
0,45
0,5
0,55
0,6
0,65
0,7
0,75
0,8
0,85
0,9
(P1 - P2)/P1
d/D ≤ 0.3
0.5
0.6
0.75

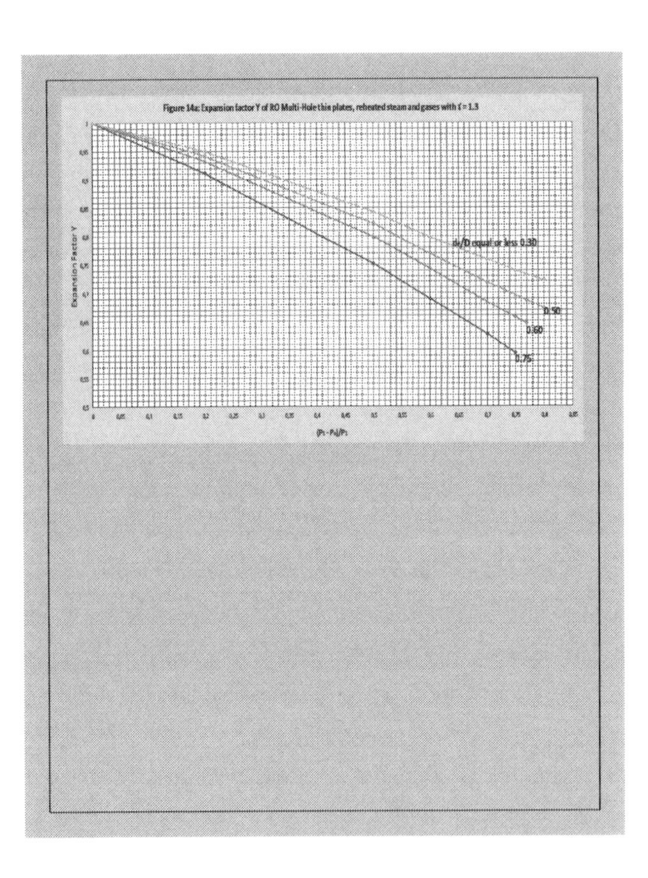

Figure 14a: Expansion factor Y of RO Multi-Hole thin plates, reheated steam and gases with κ = 1.3

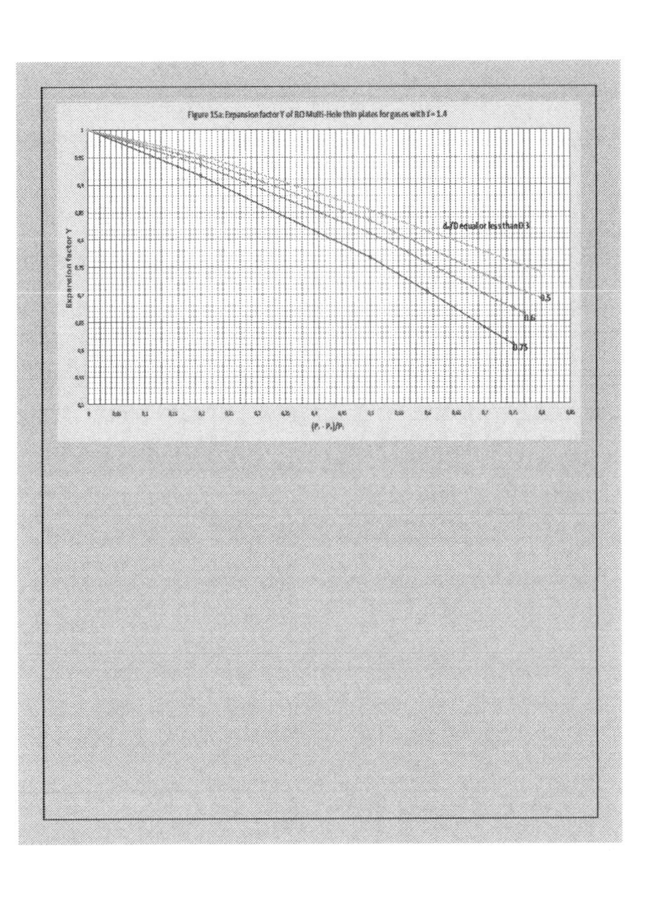
Figure 15a: Expansion factor Y of RO Multi-Hole thin plates for gases with k = 1.4
Expansion factor Y
(P₁ - P₂)/P₁
d_o/D equal or less than 0.3
0.5
0.6
0.75

4. STRUCTURAL DESIGN

The structural design of this type of ROs refers to the thickness of the plates and the thickness of the cylindrical part.

For plates with n = one hole, calculate the thickness of the plate applying the following equation:

$$\mathbf{t = [k_2((P_1 - P_F)/N)D^2/4S_M]^{0.5}} \qquad (20)$$

t = Plate thickness (mm)

P_1 = Pressure before the RO (kg/cm^2 abs.)

P_F = Pressure after the RO (kg/cm^2 abs.)

N = Number of stages

D = External diameter of the plate (mm)

S_M = Allowable stress of the plate material (kg/cm^2).

k_2 = Correction factor from the Figure 16

For plates with n > 3 holes, use the following equation:

$$\mathbf{t = [0.3((P_1 - P_F)/N)D^2R/S_M(R - d_o)]^{0.5}} \qquad (21)$$

R = Minimum straight distance between the centers of two adjacent holes (mm)

d_o = Diameter of each hole (mm)

The thickness T of the cylindrical part of the RO with plates of n = 1 hole and n > 3 holes is calculated as a pipe that has the internal pressure P_1 using the following equation:

$$T = P_1D/2(S_M + 0.4P_1) \quad (22)$$

The total length of the RO, L, if the plates are separated between them 1D and we assume that the first and last plates are also to 1D from the RO entrance and the RO outlet, is $L = 2D + (N - 1)D$.

The RO of N stages designed to pass compressible fluids must be checked also against fatigue failures in the downstream piping.

Use the Figure 17 to check if the noise provoked by the last plate is in the zone without acoustic failure. In this graph enter in the OX axis with D_F/t_F, D_F is the internal diameter of the RO downstream piping and t_F its thickness and enter in the OY axis with $M_F(P_1 - P_F)/NF$. M_F is the Mach number in the downstream piping at the RO outlet and F the pressure recovery factor of the last stage obtained in the Figure 3 or in the Figure 6.

Calculate the Mach number M_F with this equation:

$$M_F = 1.13(W/D_F^2)[v_{eF}/P_Fg]^{0.5} \quad (23)$$

The value of v_{eF} is calculated with the equation (18) as $v_{eF} = (P_1/P_F)^{1/g}v_{e1}$

If the crossing point is in the zone with pipe failures, change the design of the last stage to avoid this problem.

As the ROs with N stages have a great pressure loss it is convenient in order to avoid piping vibrations to install an anchor or a stiff support near it.

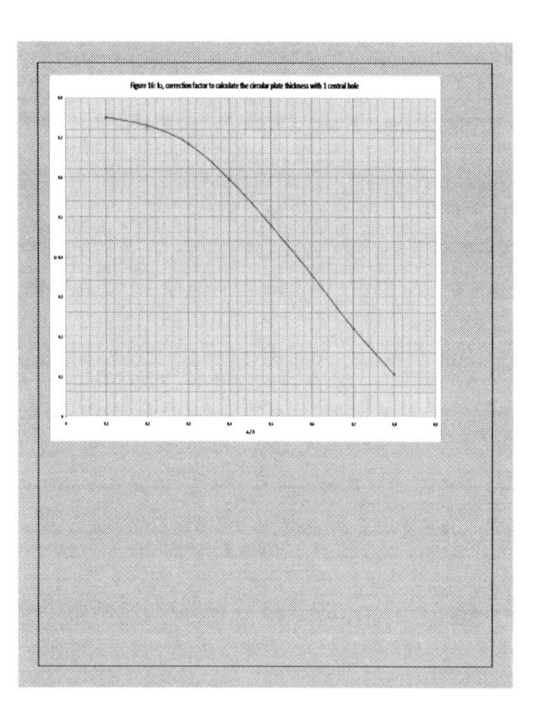

Figure 16: k_2, correction factor to calculate the circular plate thickness with 1 central hole

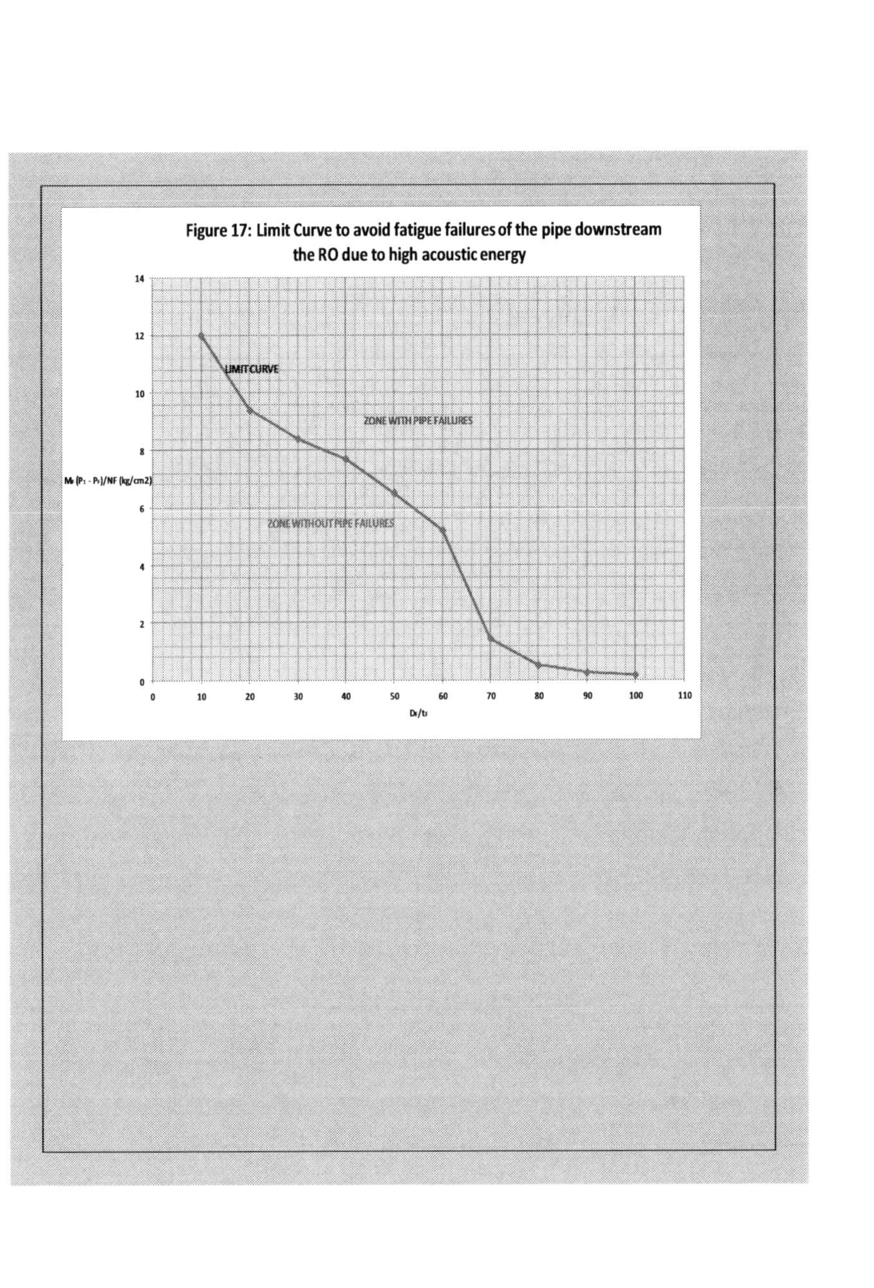

Figure 17: Limit Curve to avoid fatigue failures of the pipe downstream the RO due to high acoustic energy

5. REFERENCES

[1] Crane Technical Paper No. 410 "Flow of Fluids through Valves, Fittings and Pipe".

[2] J. P. Tullis. NUREG/CR-6031 "Cavitation Guide for Control Valves". April 1993.

[3] E. Casado. "Look at orifice plates to cut piping noise, cavitation". POWER, September 1991.

[4] E. Casado. "Avoid vibration, noise and cavitation using multi-hole plates". INGENEIRIA QUIMICA, December 1991.

[5] E. Casado. "Behavior experience of the multi-hole restriction plates". INGENIERIA QUIMICA, October 2003.

[6] E. Casado. "Calculation of ROs with N stages". INDUSTRIA QUIMICA, December 2014.

[7] M. W. Benjamin and J. G. Miller. "The Flow of Saturated Water through Throttling Orifices". Transactions of the ASME. July 1941.

[8] T. J. Rholoff and I. Catton. "Low Pressure Differential Discharge Characteristics of Saturated Liquids Passing through Orifices". Transactions of the ASME. September 1996.

[9] D. Kirk. Technical Memorandum. "Flow through Orifice Plates in Compressible Fluid Service at High Differential Pressure". December 19, 2005.

[10] F. L. Eisinger and J. T. Francis. "Acoustically Induced Structural Fatigue of Piping Systems". Journal of Pressure Vessel Technology. November 1997. Vol. 121/438-443.

[11] F. L. Eisinger. "Design Piping Systems Against Acoustically Induced Structural Fatigue". Journal of Pressure Vessel Technology. August 1997. Vol.119/379-383.

[12] Warren C. Young and Richard G. Buydnas. "Roark's Formulas for Stress and Strain". McGraw-Hill, seventh Edition.

[13] S. Timoshenko. "Strenght of Materials". Espasa-Calpe 1957.

[14] S. Timoshenko. "Vibration Problems in Engineering". D. Van Nostrand Company Tnc. May 1946.

[15] E. Casado. "Design guide for restriction orifice plates with 1 central hole". Revision 2. June 2021. Amazon.

[16] E. Casado. "Design guide for Multi-Hole RO plates with n > 3 holes. Revision 3. July 2021. Amazon

[17] P. I. PROCCESS INSTRUMENTATION. www.piprocess.com

[18] D. M. La Rosa, M. M. Rossi, G.Ferrarese, S. Malavasi. “On the pressure losses through multi-stage perforated plates”. Politecnico di Milano. Journal of Fluids Engineering. January 2021.

CHAPTER 4

LONG AND THICK RESTRICTION ORIFICES FOR LIQUIDS

LONG AND THICK RESTRICTION ORIFICES FOR LIQUIDS

Design Guide

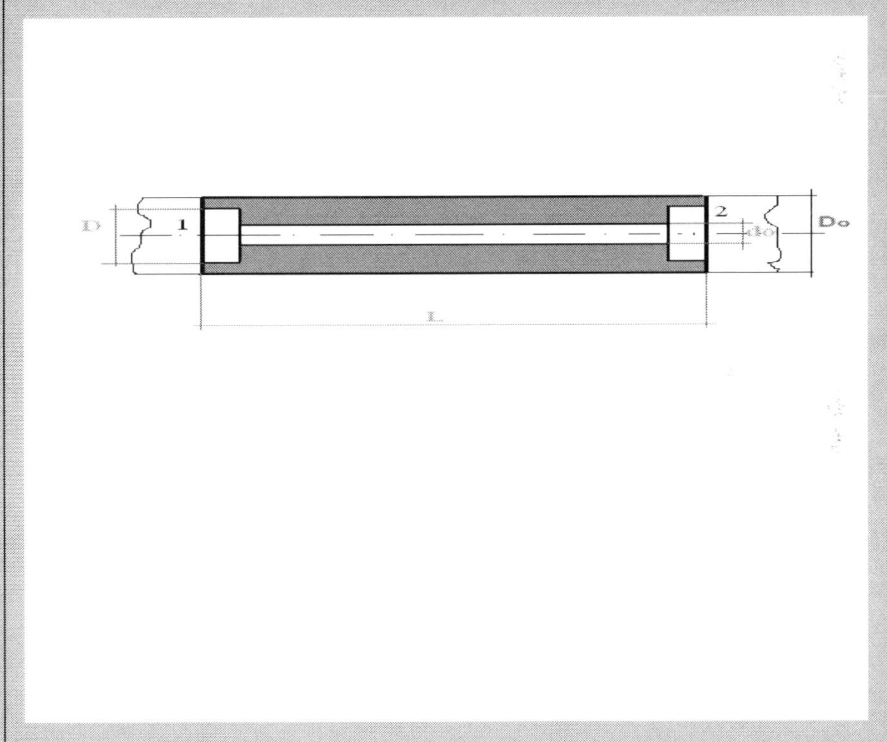

Emilio Casado

Consulting Engineer

Revision 2

June 2017

CONTENT

1. INTRODUCTION

These restriction orifices (ROs) are used when it is necessary to produce a high pressure drop with low flows. They are very different and less known as the popular RO plates.

For example, sometimes these orifices are installed in the minimum flow recirculation lines of the high-pressure pumps and the pump's manufacturer may supply in a package this type of RO as a part of the pump. Also, they are installed in the primary system of the nuclear power plants, connecting the part of nuclear class 1 with the part of nuclear class 2 in order to limit the flow of steam and water in case of pipe breaks.

This type of ROs also may be assimilated to the fuel injector nozzles of the diesel engines or rocket motors.

This document is based on the author's experience studying a big number of these orifices knowing their dimensions, the designed flow rate and the pressure drop. The author has designed and installed some long and thick orifices in the Auxiliary Feedwater System of a Spanish nuclear power plant.

The document shows how to calculate and design the ROs for liquids with cavitation or not.

2. CHARACTERISTICS OF THE LONG AND THICK ROs

The long and thick ROs have usually a ratio of L/d_o equal or greater than 5, being L the length and d_o the diameter of the hole, as shows the Figure 1.

The RO has the same external diameter as the pipe where it is installed and sometimes are named as insert ROs. Usually, the piping size is 6 inches or less.

The hole of the RO is made drilling a stainless-steel rod and the roundness of the entrance and the relative roughness of the hole has a big influence on the flow coefficient C_d

The design showed in this document corresponds to an entrance with the maximum roundness of r/d = 0.15 and up, with a resistance coefficient of 0.04 and a relative roughness of the hole, e/d_o, between 0.00005 and 0.0001. See the Reference [1].

The resistance coefficient of the RO outlet is 1, so the total resistance coefficient of the entrance plus the outlet is 1.04

For other roundness entrance as for example r/d = 0 (sharp- edge entrance) or r/d = 0.01 the corresponding values of C_d are lower than those showed in the Figure 2 and are not included in this document.

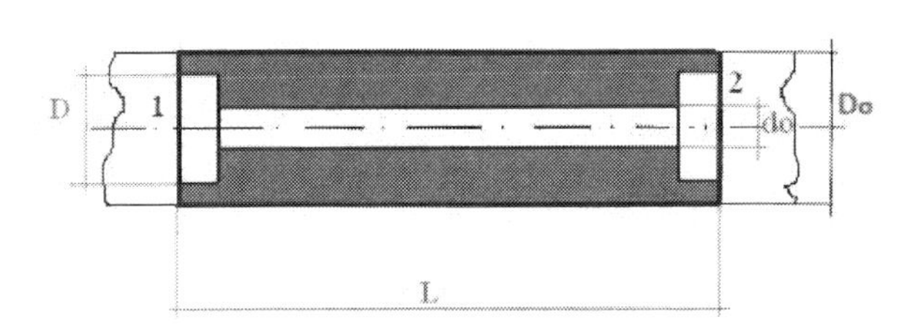

D_o = External piping diameter
D = Internal piping diameter
d_o = Orifice diameter
L = Orifice length

Figure 1

Long and thick restriction orifice (RO)

3. HOW TO DESIGN THESE ROs FOR LIQUIDS

In the References [2], [3], [4], [5], [6] and [7] is stated that the studies and experiments of fuel injector nozzles show that they have cavitation if S, the cavitation parameter, is equal or less than S_i that is the incipient cavitation. S is defined by the following equation:

$$S = (P_1 - P_V)/(P_1 - P_2) \quad (1)$$

P_1 = Pressure before the orifice (kg/cm² abs.)

P_2 = Pressure after the orifice (kg/cm² abs.)

P_V = Vapor pressure of the fluid (kg/cm² abs.)

The relation between the flow rate and the pressure drop to design the RO is given by the following equations:

$$P_1 - P_2 = 0.64Q^2/v_eC_d^2d_o^4 \quad \text{or} \quad Q = 1.25C_dd_o^2[v_e(P_1 - P_2)]^{0.5} \quad (2)$$

Q = Flow rate (m³/h)

d_o = Hole diameter (mm)

C_d = Discharge coefficient of the RO

v_e = Specific volume of the fluid at the RO entrance (m³/kg)

The numerical value of S_i depends, as it's explained in 3.2.1, of L_o/d_o

3.1 Design when there is no cavitation

When $S > S_i$, that is $P_1 - P_2 < (P_1 - P_V)/S_i$, there is no cavitation in the RO and taking into account its inlet and outlet and the length L_o of the hole diameter d_o, the RO has a total resistance coefficient, K, given by this equation:

$$\mathbf{K = 1.04 + fL_o/d_o} \qquad (3)$$

The flow coefficient $C_d = 1/K^{0.5}$ and the friction factor for the relative roughness indicated in the point 2, may be approached by the following equation: $f = 0.055/Re^{0.10}$

In the equation (3) L_o and d_o are given in mm.

The Figure 2 shows the calculated values of C_d as a function of logRe and L_o/d_o being the Reynolds number, $Re = 1.27(r/m)(Q/d_o)$, r the fluid density at the RO entrance in kg/m^3 and m the fluid dynamic viscosity at the RO entrance in kg/h-mm.

If we know the values of P_1, P_2, P_V and Q, to design the RO, calculating L_o and d_o to design a RO that provokes the pressure drop $P_1 - P_2$ for the flow rate Q, follow this process:

Select a value for d_o and obtain the Reynolds number.

With the equation (2) calculate C_d

In the Figure 2, obtain the corresponding value of L_o/d_o and from this value the length L_o

Selecting a greater value for d_o, the corresponding value of L_o will be greater and on the contrary.

If we know the values of P_1, P_R, P_V, L_o and d_o and it's necessary to calculate Q and $P_1 - P_2$, follow this process:

Select a value of Q and taking into account the downstream piping layout from the RO outlet to the end of the discharge of the piping with the pressure P_R, calculate P_2. Note that $S > S_i$ if there is not cavitation.

Calculate Re and obtain C_d in the Figure 2.

Calculate C_d with the equation (2).

If both C_d values are the same, Q and $P_1 - P_2$ are the values looked for. If not try another Q value and follow the process.

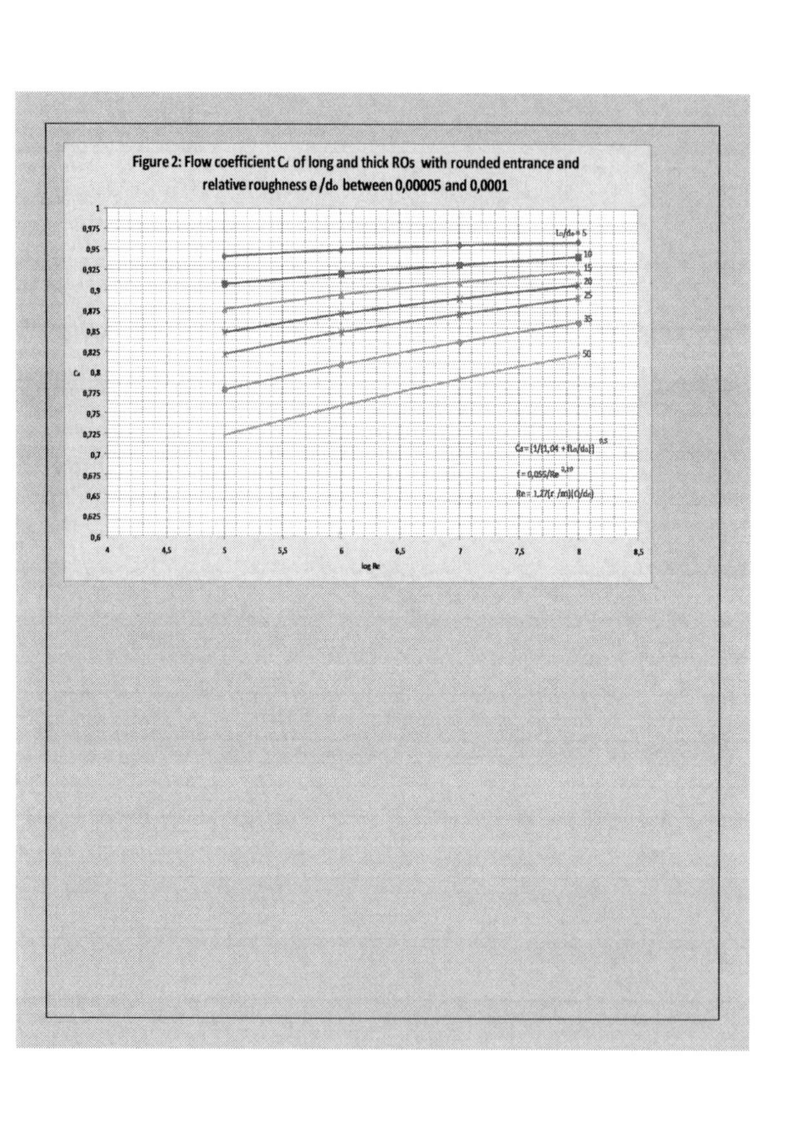

Figure 2: Flow coefficient C_d of long and thick ROs with rounded entrance and relative roughness e/d_o between 0,00005 and 0,0001

3.2 Design when there is cavitation

When S is equal or less than S_i, that is $P_1 - P_2$ is equal or greater than $(P_1 - P_V)/S_i$, there is cavitation in the RO.

The big quantity of tests, done by different researchers, shows that after the incipient cavitation begins and increases its severity, S decreases.

When the cavitation begins, the flow coefficient C_d, is independent of L_o and d_o and it depends only of S, according to this equation:

$$C_d = 0.61S^{0.5} = 0.61[(P_1 - P_V)/(P_1 - P_2)]^{0.5} \quad (4)$$

See the Reference [2] to [7].

When $S = S_i$, the RO has incipient cavitation and the flow rate Q through it is constant, if the inlet pressure P_1 is constant, for whichever value of P_2 and S equal or less than S_i

Taking into account the equations (2) and (3), the flow rate in the case of cavitation, named as Q_o, is calculated with the following equation:

$$Q_o = 0.764d_o^2[v_e(P_1 - P_V)]^{0.5} \quad (5)$$

For cold water with P_V approximately equal to 0 and $v_e = 0.001$ m^3/kg, the equation (5) becomes:

$$Q_o = 0.024d_o^2P_1^{0.5} \quad (5a)$$

In this type of ROs the cavitation is not a problem when it occurs inside the orifice and it does not reach its outlet.

When still exists cavitation in the outlet section of the orifice, we say that the RO has super-cavitation. This is the design basis of this type of ROs as fuel injector nozzles, because it's necessary that they have super-cavitation in order to help the fuel atomization. See again the References [2] to [7].

But in this document, we design the ROs only as restriction orifices and we must avoid the super-cavitation to protect the orifice downstream piping from the erosion.

The equations (5) and (5a) give the maximum flow rate Q_o that may pass through the RO and corresponds to the beginning of the cavitation. For values of $S < S_i$ (higher cavitation levels) the value of Q_o is constant and it doesn't depend of $P_1 - P_2$.

The Figure 3 gives directly Q_o for water at 20 ºC.

So, if the initial known data are Q_o and $P_1 - P_2$, when there is cavitation the orifice diameter d_o is calculated directly with the equations (5) or (5a).

Note that a conclusion of the running of the RO with cavitation is that the flow rate is constant though P_2 changes, provided S is equal or less than S_i

It's like a control valve that maintains the flow constant.

But as it's said before, to avoid damages in the RO downstream piping, it's necessary to avoid the super-cavitation, designing the RO with enough length to maintain the cavitation inside it, as it is explained in 3.2.2 and done with a cavitation resistant material as it is said in 4.

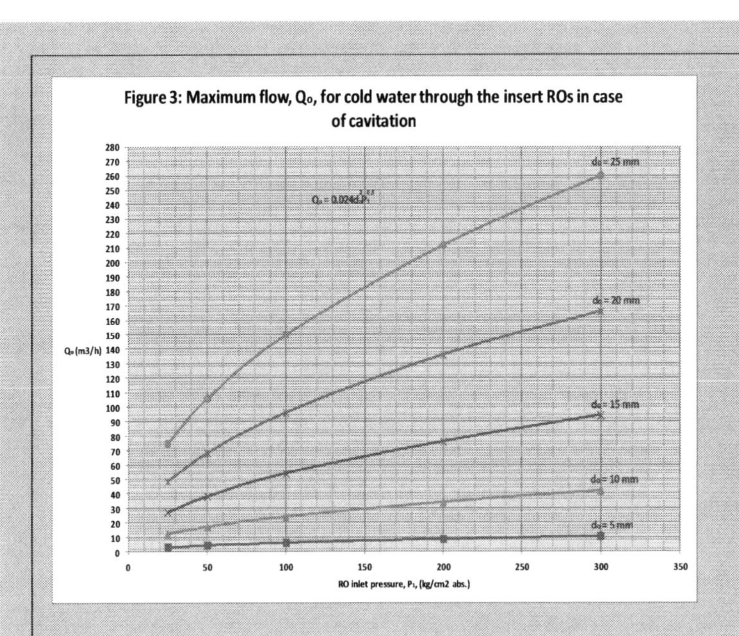

Figure 3: Maximum flow, Q_o, for cold water through the insert ROs in case of cavitation

3.2.1 The incipient cavitation coefficient S_i

We can consider that when $S = S_i$ all the equations for no cavitation and for cavitation, showed before, apply simultaneously.

Taking as reference cold water at 20 ºC with m = 0.0036 kg/h-mm and $v_e = 0.001\ m^3/kg$ the equation that gives S_i for a long and thick RO with rounded entrance and e/d_o = 0.00005 to 0.0001 is:

$$S_i = 2.7/[\,1.04 + (0.022/d_o^{0.1}P_1^{0.05})(L_o/d_o)] \qquad (6)$$

In this equation, basically S_i depends only of L_o/d_o because the variation with d_o and P_1 is very small.

The curve of the Figure 4 shows the theoretical values of S_i as function of L_o/d_o

This curve gives the minimum value of L_o/d_o that must have the RO for every value of $S = S_i$ in order to avoid the cavitation. It's possible to simplify the equation (6) and give directly L_o/d_o by this other equation that conservatively gives a little higher value:

$$L_o/d_o = 232\log(2.54/S_i) \qquad (7)$$

See the References [3] and [6] that include tests and experiments with values of S_i from 2 to 1.5

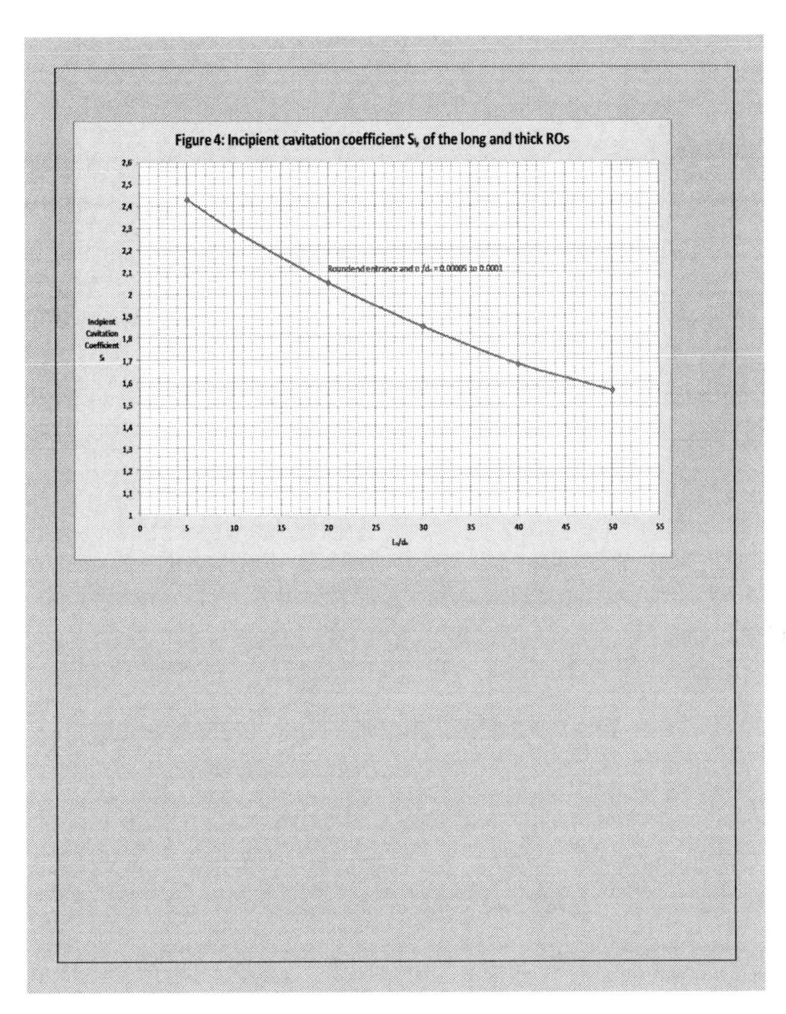
Figure 4: Incipient cavitation coefficient S_i, of the long and thick ROs
2,6
2,5
2,4
2,3
2,2
2,1
2
1,9
1,8
1,7
1,6
1,5
1,4
1,3
1,2
1,1
1
Incipient Cavitation Coefficient S_i
0
5
10
15
20
25
30
35
40
45
50
55
L_o/d_o
Roundend entrance and ε /d_o = 0.00005 to 0.0001

3.2.2 The RO length to avoid the super-cavitation

As it's said before, the super-cavitation in a thick and long RO occurs when the cavitation length reaches the RO outlet.

Taking into account the results of the tests of the References [2] to [7] and the design of different ROs of this type installed in some nuclear power plants, it's possible to obtain conservatively which is the minimum L_o/d_o value that must have the RO to maintain the cavitation inside it, avoiding the super-cavitation.

As show the tests, the cavitation begins at the RO inlet and extends through the RO length L_o a distance L_c. If $L_o > L_c$ the RO has not super-cavitation.

The Figure 5 gives the minimum values of L_o/d_o to avoid super-cavitation for ROs with rounded or sharp edge entrances and relative roughness of 0.00005 to 0.001, but it's always recommendable to design the RO entrance with the maximum possible roundness.

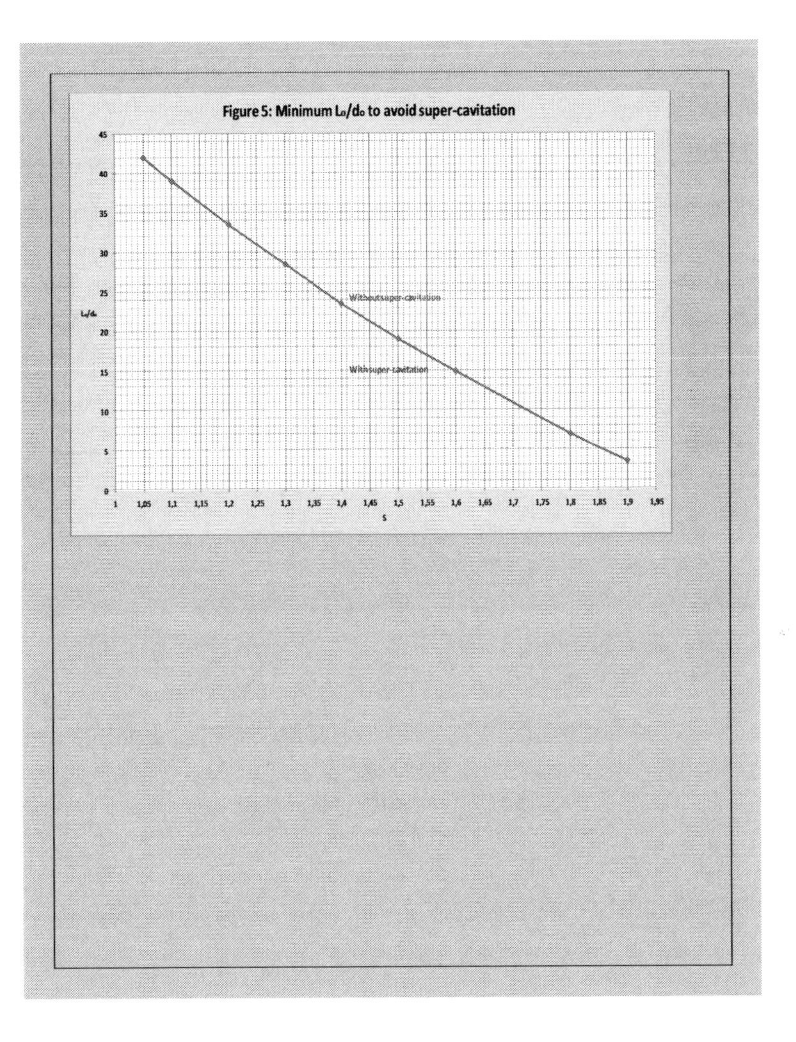

Figure 5: Minimum L_o/d_o to avoid super-cavitation

4. STRUCTURAL DESIGN

The minimum thickness of the RO, $t_o = (D_o - D)/2$ according to the Figure 1 is calculated with the following equation:

$$t_o = P_1 D / 2(S_M + 0.4P_1) \qquad (8)$$

t_o = Thickness (mm)

P_1 = Pressure at the entrance of the RO (kg/cm^2)

D = Internal diameter of the connecting ends to the piping (mm)

S_M = Allowable stress of the RO material (kg/cm^2)

The thickness in the rest of the RO, $t = (D_o - d_o)/2$ will be greater than t_o.

The RO material must be stainless steel or alloyed steel in order to avoid damages inside it if there is cavitation.

Due that these ROs provoke high pressure drops, it's advisable to install an anchor or a stiff support in the piping, near the RO, in order to avoid piping movements.

5. NOMENCLATURE

C_d = Flow coefficient

D = Internal piping diameter (mm)

d_o = RO hole diameter (mm)

D_o = External piping diameter (mm)

f = Friction factor

K = Resistance coefficient

L_o = Length of the hole (mm)

L_c = Cavitation length (mm)

L = RO length (mm)

P_1 = Pressure at the RO inlet (kg/cm^2 abs.)

P_2 = Pressure at the RO outlet (kg/cm^2 abs.)

P_V = Fluid vapor pressure (kg/cm^2 abs.)

P_R =Ambient pressure where discharges the RO downstream piping (kg/cm^2 abs.)

Q = Flow rate through the RO (m^3/h)

Q_o = Maximum flow rate in case of cavitation (m^3/h)

S_i = Incipient cavitation

v_e = Specific volume of the fluid (m^3/kg)

W = Flow rate through the RO (kg/h)

r = Fluid density (kg/m^3)

m = Fluid viscosity (kg/h-mm)

e/d_o = Relative roughness of the hole

6. REFERENCES

[1] Crane Technical Paper No. 410. "Flow of Fluids Through Valves, Fittings and Pipe".

[2] H. Chaves and C. H. Ludwig. "Characterization of Cavitation in Transparent Nozzles Depending on the Nozzle Geometry". Department of Mechanics and Fluid Dynamics. Freiberg University, Germany. Proceedings of the 20th ILASS-Europe Meeting 2005.

[3] W. H. Nurick, T. Ohanian, D. G. Talley and P. A. Strakey. "Impact of L/D on 90 Degree Sharp-Edge Orifice Flow with Manifold Passage Cross Flow". Air Force Research Laboratory. April 30, 2007.

[4] K. Sato and Y. Saito. "Unstable Cavitation Behavior in a Circular-Cylindrical Orifice Flow". Department of Mechanical and Mechanical Systems Engineering. Kanazawa Institute of Technology. Nonoichi-machi, Ishikawa 921-8501, Japan. 2001.

[5] B. Ebrahimi, G. He, Y. Tang, M. Franchek, D. Lin, J. Pickett, F. Springett and D. Franklin. "Characterization of high-pressure cavitating flow through a thick orifice plate in a pipe of constant cross section". International Journal of Thermal Sciences 114(2017) 229-240.

[6] David P. Schmidt. "Cavitation in Diesel Fuel Injector Nozzles". University of Wisconsin. Madison 1997.

[7] D. P. Schmidt, T. F. Su, K. H. Goney, P. V. Farrell and M. L. Corradini. "Detection of Cavitation in Diesel Fuel Injector Nozzles". Engine Research Center. University of Wisconsin. Madison, December 2014.

[8] E. Casado. "Restriction orifices design for high pressure loss". Ingeniería Química. July/August 1994.

CHAPTER 5

LONG AND THICK RESTRICTION ORIFICES FOR SATURATED WATER, STEAM (SATURATED AND REHEATED) AND GASES

LONG AND THICK ORIFICES FOR SATURATED WATER, STEAM (SATURATED AND REHEATED) AND GASES.

Design Guide

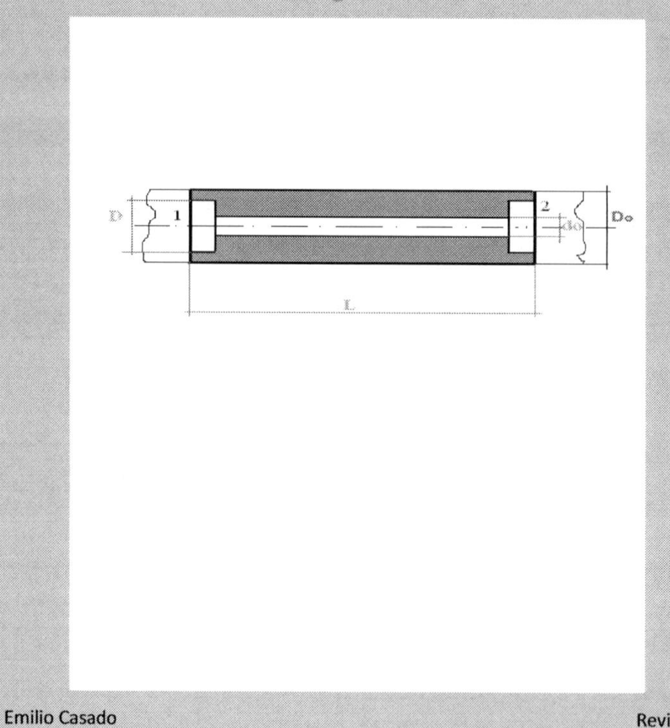

Emilio Casado

Consulting Engineer

Revision 2

April 2018

CONTENT

1. INTRODUCTION

This type of Restriction orifice, RO, designed to discharge saturated water, steam or gases are installed in the nuclear power plants, for example in some connections of pipes to the reactor coolant system to limit the flow of water and steam in case of breaks and in the reactor vessel vent lines to discharge H_2. In this document we name it as insert ROs because they are installed in the piping with the same outer diameter as if were inserted in it.

Normally they are calculated as nozzles because there is not much published technical literature about its design guides.

In this Guide, the author presents its design assuming the RO as a model formed by a pipe of length L and internal diameter d_o plus an entrance and an exit. The resistance coefficient of the entrance plus the exit will be 1.04 for a rounded entrance and the relative roughness of the pipe will be between 0.00005 and 0.0001 so the friction factor f may be approached by $f = 0.055/Re^{0.1}$ and the Reynolds number is $Re = 1.27rQ/ md_o$. See the Reference [5]

The resistance coefficient of the RO, K, will be:

$$\mathbf{K = 1.04 + fL_o/d_o} \qquad (1)$$

The dimensions of L and d_o are mm.

The Figure 1 shows a sketch of this type of RO.

The flow coefficient of the RO, C_d, is given in the Figure 2 and taken from the Reference [5]. The relation between K and C_d is $K = 1/C_d^2$

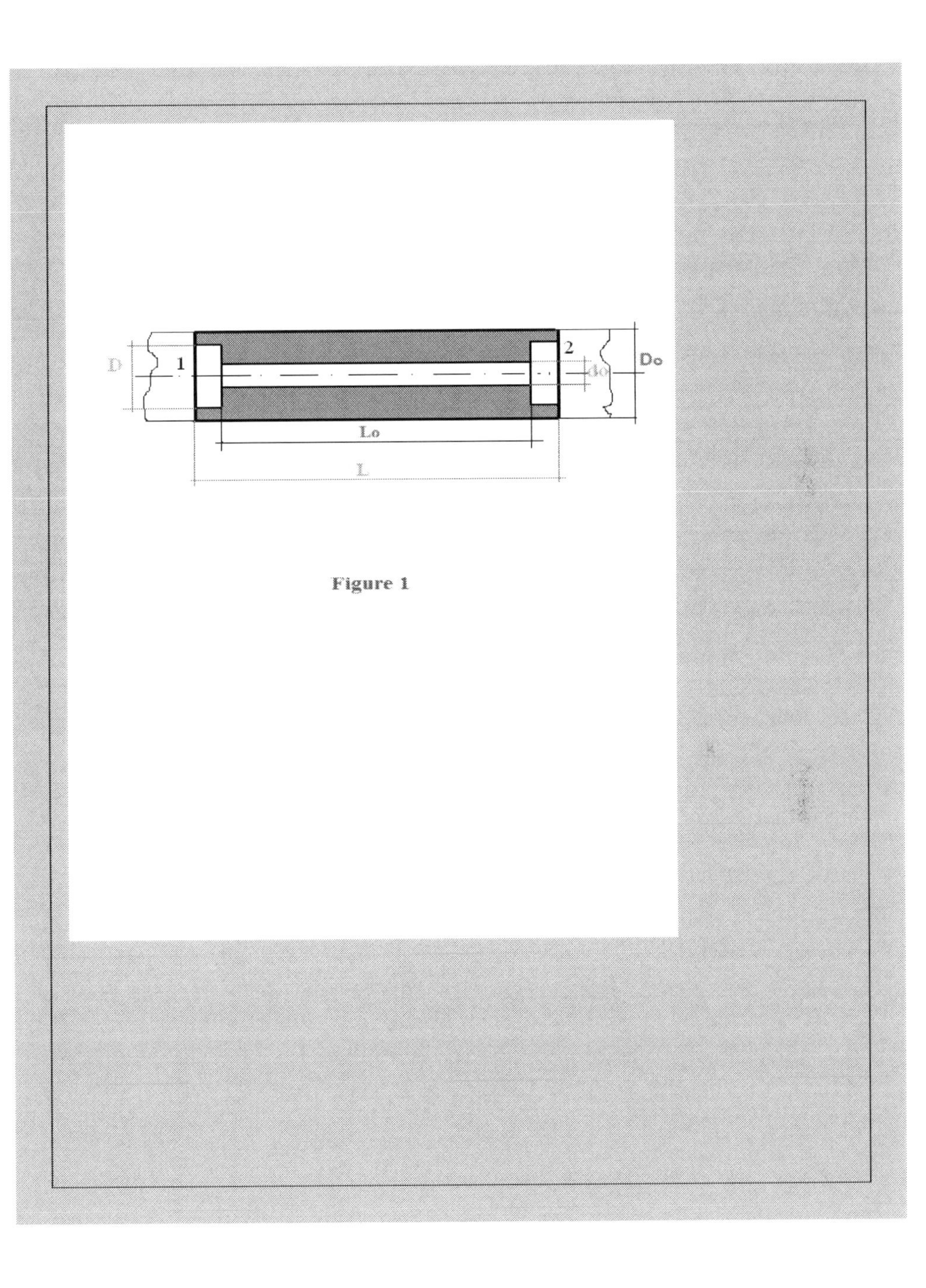

Figure 1

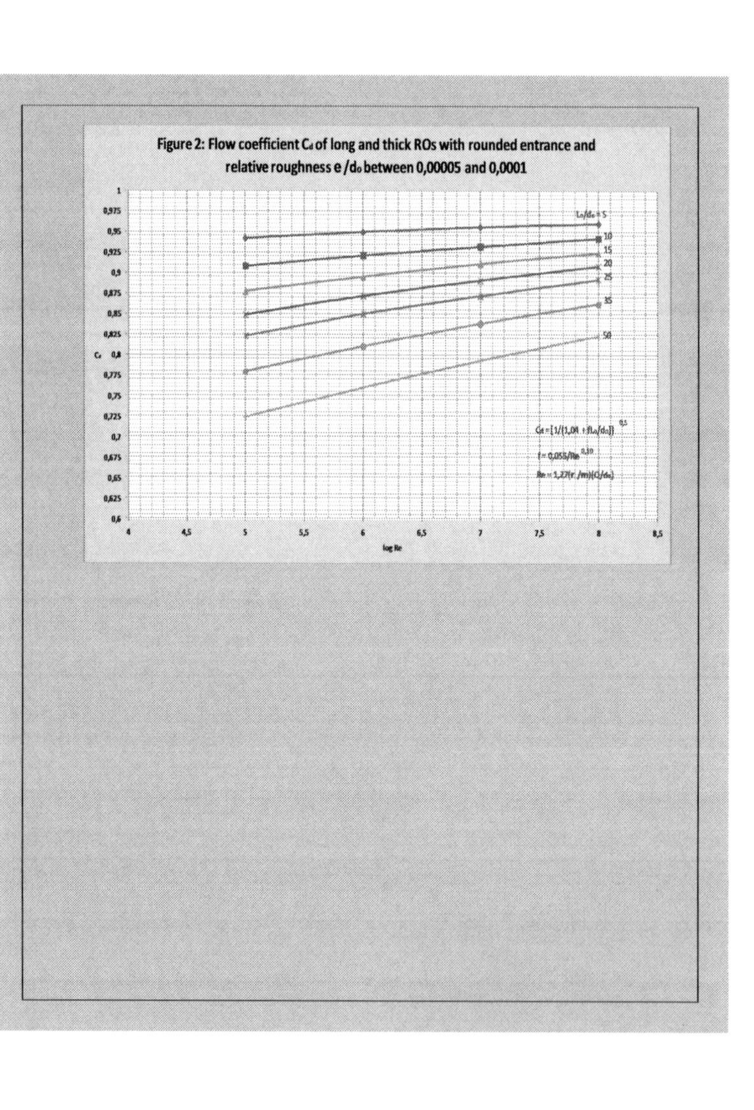

Figure 2: Flow coefficient Cd of long and thick ROs with rounded entrance and relative roughness e /do between 0,00005 and 0,0001
1
0,975
0,95
0,925
0,9
0,875
0,85
0,825
Cd
0,8
0,775
0,75
0,725
0,7
0,675
0,65
0,625
0,6
4
4,5
5
5,5
6
6,5
7
7,5
8
8,5
log Re
Lo/do = 5
10
15
20
25
35
50

2. DESIGN OF THE RO INSERT FOR SATURATED WATER

When the fluid is saturated water the relation between the flow rate and the pressure drop through the pipe is given by the following equation developed by the author and exposed in the Reference [1]:

$$\mathbf{W = 0.19C_Ed_o^2[(P_S - P_2)/v_{e1}K]^{0.5}(2.65 + lnP_S)} \qquad (2)$$

W = Flow rate (kg/h)

C_E = Expansion coefficient of the saturated water taken from the Figure 3

d_o = Internal diameter of the orifice (mm)

P_S = Saturation pressure at the RO inlet (kg/cm^2 abs.)

P_2 = Pressure at the RO outlet (kg/cm^2 abs.)

v_{e1} = Specific volume of the saturated water at the RO inlet (m^3/kg)

K = Resistance coefficient of the RO given by the equation (1)

In the Figure 3 to obtain C_E enter in the graph with DP/P_S equal to $(P_S - P_2)/P_S$

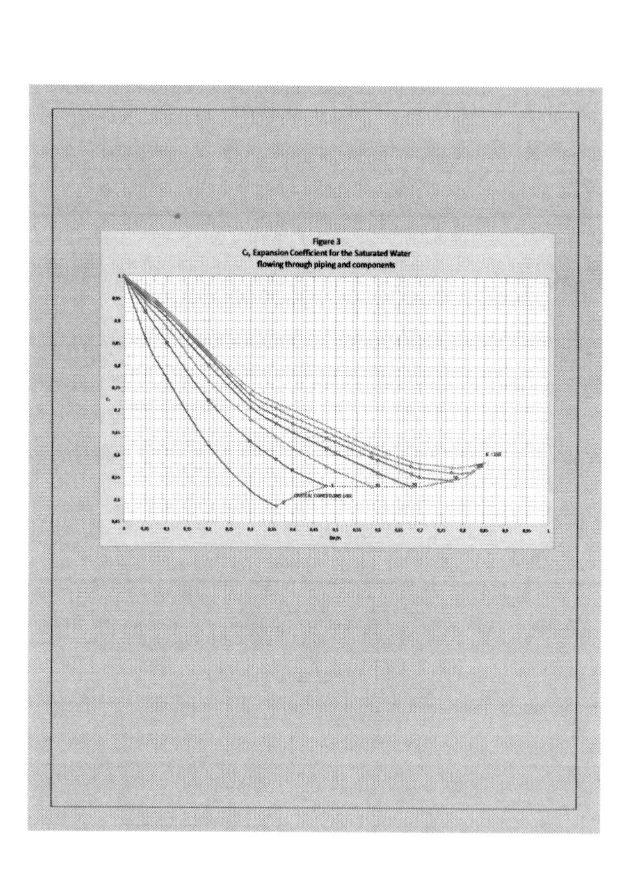

Figure 3
C_e, Expansion Coefficient for the Saturated Water flowing through piping and components

This type of RO produces a high pressure drop and usually the pressure P_2 corresponds to the critical pressure P_{CR}. In these cases, take directly C_E in the Figure 3 for the end of the K value in the critical conditions line.

When P_S is equal or less than 17.5 kg/cm^2 abs. and DP/P_S equal or less than 0.15, correct the flow rate calculated with the equation (2) multiplying W by the correction factor C_1 given by the following equation:

$$C_1 = 0.95K^{0.05} \qquad (3)$$

When as it is usually there are critical conditions at the RO outlet,

$P_2 = P_{CR}$, it is possible to obtain directly P_{CR} in the Figure 4 and as a function of P_S and K.

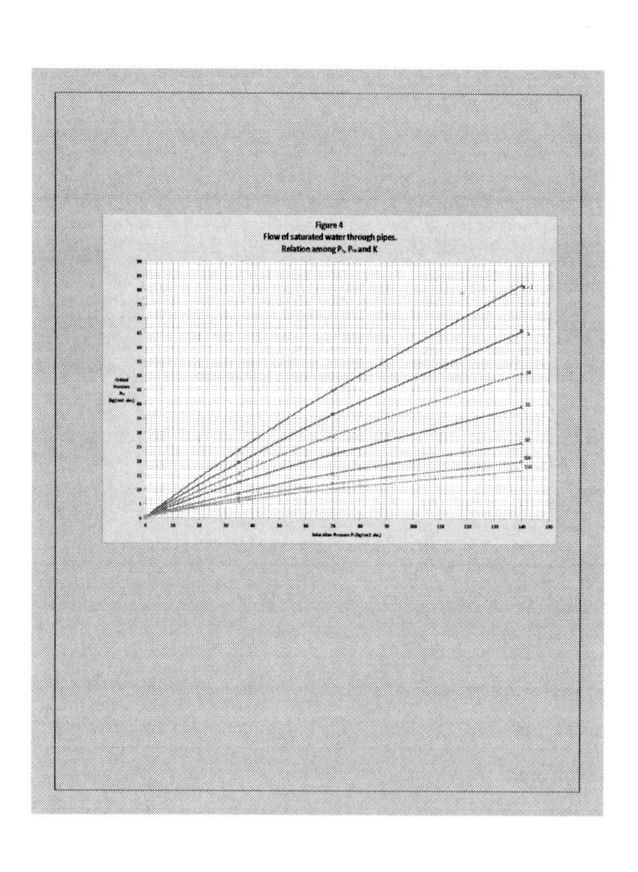
Figure 4
Flow of saturated water through pipes.
Relation among P_s, P_m and K

At the RO outlet part of the saturated water has been converted to steam, and the specific volume of the mix, v_{e2}, is higher than v_e, the specific volume of the saturated water at the RO inlet, so the corresponding velocity of the fluid tends to be high.

To protect the pipe downstream the RO is convenient to limit this velocity. If the material of the pipe is carbon steel a limit of 25 m/s for continuous service and 30 m/s for intermittent service, could be valid. For higher velocity values, select alloy steel or stainless steel as the material of the downstream piping.

The velocity of the fluid in the downstream pipe is determined by the following equation:

$$\mathbf{V_2 = 3.54 v_{e2} W/D^2} \qquad (4)$$

V_2 = Fluid velocity in the pipe at the RO outlet (m/s)

v_{e2} = Specific volume of the fluid at the RO outlet (m^3/kg)

W = Flow rate (kg/h)

D = Internal diameter of the RO downstream pipe (mm)

To obtain v_{e2} it is necessary to calculate the quantity of steam (x %) produced in the expansion from P_S to P_2 in the RO. This may be done assuming an isentropic expansion and using the Figure 5 to obtain x.

In the Steam Tables with the saturation pressure P_2 and x, calculate v_{e2}.

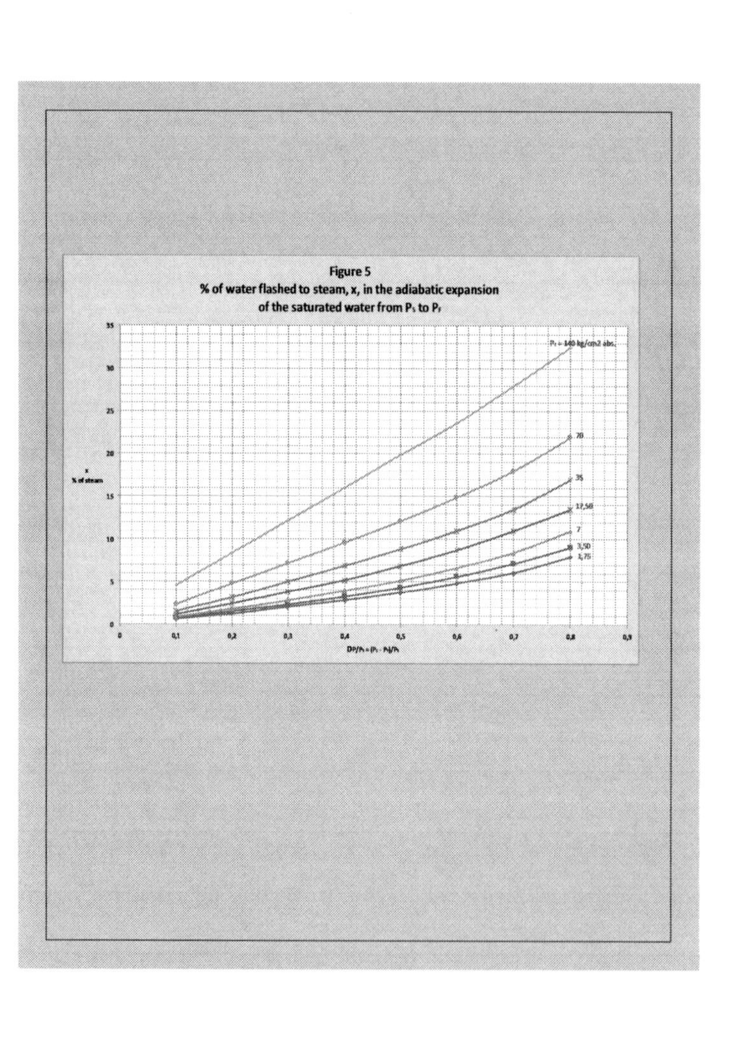
Figure 5
% of water flashed to steam, x, in the adiabatic expansion
of the saturated water from Ps to Pf
x
% of steam
35
30
25
20
15
10
5
0
0
0,1
0,2
0,3
0,4
0,5
0,6
0,7
0,8
0,9
kg/cm2 abs.
70
35
17,50
7
3,50
1,75

The calculation process may present these two scenarios:

1) The known data are P_s, P_2 and Q and it is necessary calculate L_o and d_o

2) The known data are P_s, L_o and d_o and it is necessary calculate P_2 and Q

In the first scenario as $P_2 = P_{CR}$ obtain K from the Figure 4 and obtain C_E from the Figure 3 in the critical conditions line.

Apply the equation (1) to calculate d_o

Finally calculate Re and f as it is exposed in the point 1 and applying the equation (1) obtain L_o

In the second scenario, assume a value for Q and calculate Re. Obtain C_d in the Figure 2 and $K = 1/C_d^2$

With K and P_s obtain $P_2 = P_{CR}$ in the Figure 4 and C_E in the Figure 3.

Finally apply the equation (2) and calculate W. As W = rQ, if Q is the same as the assumed value, P_2 and Q are the values looked for, if not assume other value for Q and repeat the process.

3. DESIGN THE RO INSERT FOR STEAM (SATURATED AND REHEATED) AND GASES

The equation that relates the flow rate and the pressure drop is:

$$W = 1.25d_o^2Y[(P_1 - P_2)/v_eK]^{0.5} \quad (5)$$

W = Flow rate (kg/h)

d_o = Orifice internal diameter (mm)

Y = Expansion factor (Figure 6 and 7)

P_1 = Pressure of the fluid at the RO inlet (kg/cm^2 abs.)

$P_2 = P_{CR}$ = Steam critical pressure after the RO (kg/cm^2 abs.)

v_e = Specific volume of the fluid before the RO (m^3/kg)

K = Resistance coefficient of the RO

Note: This equation is taken from the Reference [2].

Y is taken from the Figure 6 or 7 with $DP = P_1 - P_2$. The Figure 6 applies to the steam, saturated or reheated and gases with $a = c_p/c_v = 1.3$ and the Figure 7 to gases that have a = 1.4. Y is similar to C_E of the point 2 before.

As in the point 1, $K = 1/C_d^2$ and also applies the equation (1).

Usually there are critical conditions at the RO outlet so $P_2 = P_{CR}$. In this case the relation among P_S, P_{CR} and K is given by the Figure 8 for a = 1.3 and the Figure 9 for a = 1.4. Note that in these Figures 8 and 9, the pressures are in psia instead in kg/cm^2 abs.

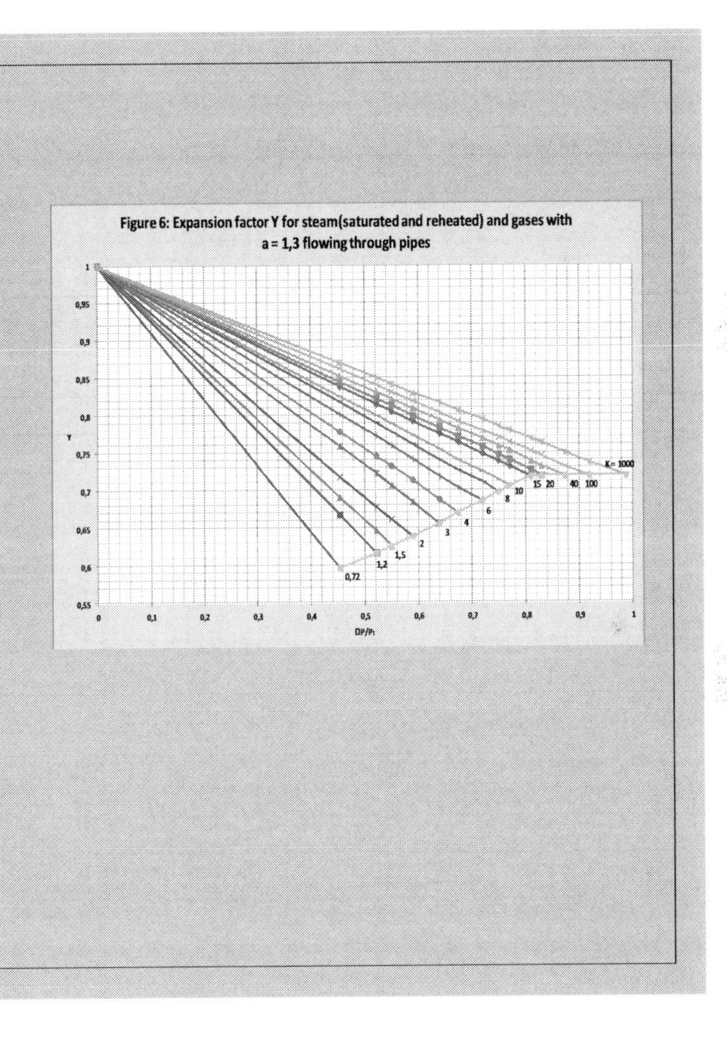

Figure 6: Expansion factor Y for steam(saturated and reheated) and gases with a = 1,3 flowing through pipes

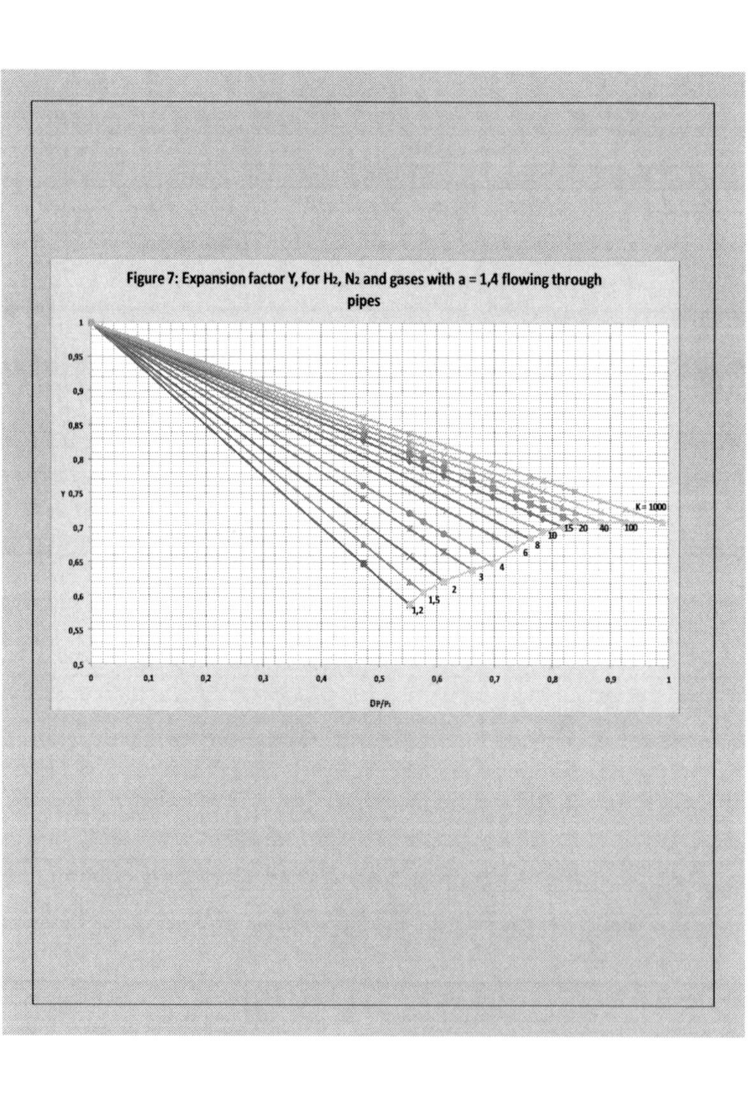
Figure 7: Expansion factor Y, for H2, N2 and gases with a = 1,4 flowing through pipes
Y
1
0,95
0,9
0,85
0,8
0,75
0,7
0,65
0,6
0,55
0,5
0
0,1
0,2
0,3
0,4
0,5
0,6
0,7
0,8
0,9
1
DP/P1
K = 1000
100
40
20
15
10
8
6
4
3
2
1,5
1,2

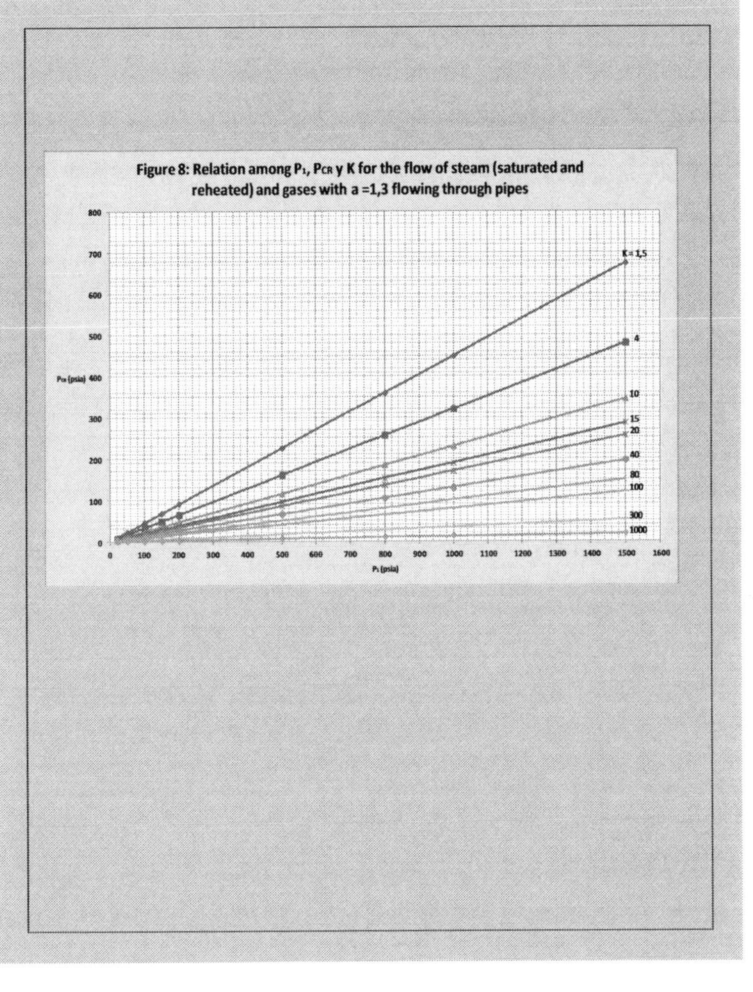

Figure 8: Relation among P_1, P_{CR} y K for the flow of steam (saturated and reheated) and gases with a =1,3 flowing through pipes

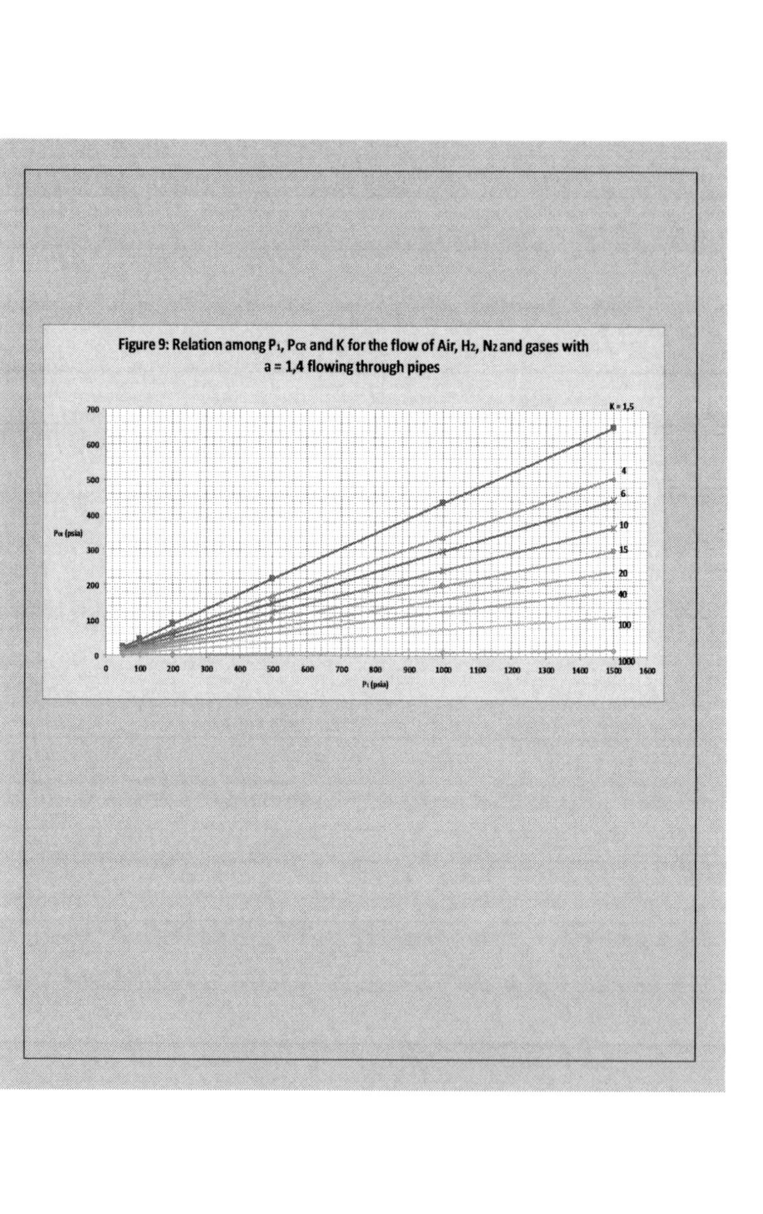
Figure 9: Relation among P_1, P_{CR} and K for the flow of Air, H_2, N_2 and gases with a = 1,4 flowing through pipes
K = 1,5
4
6
10
15
20
40
100
1000
P_{CR} (psia)
P_1 (psia)
0
100
200
300
400
500
600
700
800
900
1000
1100
1200
1300
1400
1500
1600

The calculation scenarios may be the same as those exposed in the point 2 for saturated water.

When the RO has been designed, it is necessary to check that it does not provoke acoustic fatigue failures in the downstream piping.

To check this, use the graph of the Figure 10. Enter in the graph with D_2/t_2, being D_2 the downstream piping internal diameter and t_2 its thickness, and with M_2DP, being M_2 the fluid Mach Number at the RO outlet calculated with the equation:

$$\mathbf{M_2 = 1.06(W/D_2^2)(v_{e2}/P_2)^{0.5}} \qquad (6)$$

W = Flow rate (kg/h)

v_{e2} = Specific volume of the fluid at the RO outlet (m^3/kg)

$P_2 = P_{CR}$ (kg/cm^2 abs.)

Calculate v_{e2} using this equation: $v_{e2} = v_e(P_S/P_{CR})^{1/a}$

DP is the pressure drop in the RO, $P_1 - P_{CR}$

If in the Figure 10 if the crossing point is above the limit curve, we must install some meters of the downstream piping with more thickness or increase D_2 in order to decrease the acoustic fatigue below the limit curve.

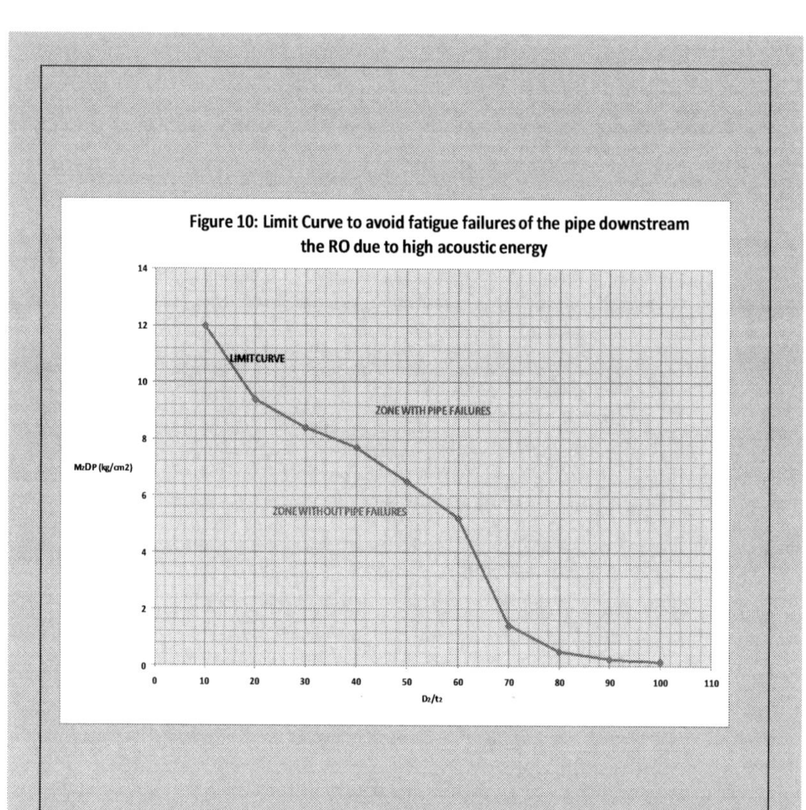

Figure 10: Limit Curve to avoid fatigue failures of the pipe downstream the RO due to high acoustic energy

4. STRUCTURAL DESIGN

The minimum thickness of the RO, $t_o = (D_o - D)/2$, see the Figure 1, is calculated with the following equation:

$$t_o = P_1 D/2(S_M + 0.4P_1) \qquad (7)$$

t_o = Minimum thickness (mm)

P_1 = Pressure at the RO inlet (kg/cm^2 abs.)

D = Internal diameter of the connecting RO ends to the piping (mm)

S_M = Allowable stress of the RO material (kg/cm^2)

The RO material must be stainless steel or alloyed steel in order to resist the cavitation effects.

Due that, these ROs provoke high pressure drop, it is convenient to install an anchor or a stiff support in the piping near the RO, in order to avoid piping movements.

5. REFERENCES

[1] E. Casado. "Improve pressure drop calculations for saturated water lines". HYDROCARBON PROCESSING. August 2013.

[2] Crane Technical Paper No. 410. "Flow of fluids through valves, fittings and pipe".

[3] F. L. Eisinger and J. T. Francis. "Acoustically Induced Structural Fatigue of Piping Systems". Journal of Pressure Vessel Technology.November 1997. Vol.121/438-443.

[4] F. L. Eisinger. "Design Piping Systems Against Acoustically Induced Structural Fatigue". Journal of Pressure Vessel Technology. August 1997. Vol. 119/379-383.

[5] E. Casado. "Long and Thick Orifices for Liquids". Design Guide. Research Gate. September 2017

CHAPTER 6

ROTARY MULTI-HOLE RO PLATES

DESIGN CHARACTERISTICS OF THE ROTARY MULTI-HOLE RO PLATES

Emilio Casado
Consulting Engineer

Revision 1
August 2021

CONTENT

1. INTRODUCTION

The rotary multi-hole plates, named by the author as EMA-1, is a combination of a butterfly valve and a multi-hole restriction orifice plate. It has very low cavitation coefficients that can decrease to values close to 1.05 or less, similar to those of the complex and expensive multi-stage pressure drop trims that tip the cavitation control valves.

These devices may consider them as multi-hole restriction plates that can regulates its position between 0º (vertical position) and 90º (horizontal position) or butterfly valves with the disc drilled by holes.

The author has developed these devices from the analysis, design and tests of different butterfly valves and multi-hole restriction orifice plates as show the References [1], [3], [4], [5] and [6]. Two prototypes of them are installed and running in a Spanish nuclear power plant.

This document exposes the flow coefficients, the incipient cavitation coefficients and the number and diameter of the holes of these devices.

In conclusion, the main advantages of the rotary multi-hole plates are:

- The extremely low incipient cavitation coefficient
- The adjusting possibility to vary the flow rate and pressure drop through them

2. THE INCIPIENT CAVITATION COEFFICIENT OF THE BUTTERFLY VALVES AND THE MULTI-HOLE RO PLATES

The incipient cavitation of a device as a valve or a restriction orifice that limit the flow rate and provoking a pressure drop corresponds to the beginning of the typical cavitation noise that sounds as little stones flowing with the fluid through the pipe. If in this point the pressure drop increases, the noise increases and therefore the cavitation. If the pressure drop decreases, the noise disappears.

When the cavitation noise begins, the relation among P_1, P_2 and P_v given by the following equation (1), is the incipient cavitation coefficient of the device C_i

$$C_i = (P_1 - P_v) / (P_1 - P_2) \qquad (1)$$

P_1 and P_2 are the fluid pressures before and after the device and P_v the fluid vapor pressure.

The C_i values of the valves and the restriction orifices come from tests. For each value of P_1, test the devices for different values of $P_1 - P_2$ and obtain the values of C_i when the incipient noise begins.
To know if one of these devices has cavitation or not when it is in operation, measure the values of P_1 and P_2 and with the value of P_v calculate the cavitation index s with the same equation (1). If $s > C_i$ there is no cavitation, if $s = C_i$ the incipient cavitation begins and finally, if $s < C_i$ there is cavitation and may corresponds to others levels as severe and damage.
The Figure 1 shows the C_i values of the standard butterfly valves, based in the References [1] and [2] and the Figure [2] shows the C_i values of the multi-hole restriction orifice plates, that come from tests developed by the author as say the References [3], [4] and [5].

3. THE ROTARY MULTI-HOLE RO PLATES (EMA-1)

Combining a butterfly valve with a multi-hole RO plate, the result is a rotary multi-hole plate, designed by the author as shows the Figures 3 and 4. This device is a multi-hole restriction orifice (RO) plate that may change Its position or a butterfly valve with the disc drilled by holes that cannot isolate the flow.

Its main characteristic is that has an extremely low incipient cavitation coefficient.

The name of this type of restriction orifice is EMA-1.

3.1 The flow coefficient C_{vo}

As shows the Figure 4, in a disc position with an opening angle a greater than 0º the total flow Q divides between the flow through the holes Q_{RO} and the flow through the disc edge Q_V.

The pressure drop DP in this device is the same through the holes than through the disc edge, therefore the flow coefficient C_{vo} will be:

$$\mathbf{C_{vo} = 0.037Q / (v_e DP)^{0.5}} \qquad (2)$$

The pressure drop through the edge disc is:

$$\mathbf{DP = (0.00137/v_e)\,(Q_V/C_v)^2} \qquad (3)$$

The pressure drop through the holes is:

$$\mathbf{DP = (0.64F/v_e)\,(Q_{RO}/Cd_e^2\cos^2 a)^2} \qquad (4)$$

Finally, the flow coefficient is:

$$\mathbf{C_{vo} = C_v + Cd_e^2\cos^2 a/21.5F^{0.5}} \qquad (5)$$

C is the discharge coefficient of the multi-hole RO plates taken from the Figure 5 and d_e the equivalent diameter of the holes.

F is the pressure recovery factor of the multi-hole RO plates and $F^{0.5}$ comes from the Figure 6.

C_v is the flow coefficient of the plate edge, similar to that of a butterfly valve. Use the approach equation (6) to estimate its value for opening angles a, equal or less than 70º, with D in mm:

$$\mathbf{C_v = 0.15x10^{-6}D^2\,(a^3 - 6.25a^2 + 412.5a)} \qquad (6)$$

Therefore, the values of C_{vo} of a rotary multi-hole plate, depends on the disc plate size, the number and diameter of the holes and the opening angle.

As an example, see the Figures 7 and 8 that show the C_{vo} values of different rotating multi-hole plates.

3.2 The incipient cavitation coefficient C_{io}

For each opening angle, the incipient cavitation coefficient of the rotary multi-hole plates has two parts. One part is the incipient cavitation coefficient of the plate edge, named as C_{iV} and expressed by the equation (7) and the other part is the incipient cavitation coefficient of the holes, named as C_{iRO} and expressed by the equation (8).

$$\mathbf{C_{iV} = 1 + 0.53(P_1 - P_v)^{0.3}(C_{iB} - 1)[21.50C_vF^{0.5}/(Cd_e\cos^2 a + 21.50C_vF^{0.5})]^2} \quad (7)$$

$$\mathbf{C_{iRO} = 1 + (C_I - 1)[Cd_e^2\cos^2 a/(Cd_e\cos^2 a + 21.50C_vF^{0.5})]^2} \quad (8)$$

The incipient cavitation coefficient of the rotary multi-hole plate, C_{io} is the great of both. The minimum value of C_{io} corresponds to the opening angle a_o where the values of C_{iV} and C_{iRO} are the same. For $a < a_o$ the C_{iV} values of the plate edge are greater than the C_{iRO} of the plate holes and for $a > a_o$ is on the contrary. Therefore, $C_{io} = C_{iV}$ when $a < a_o$ and $C_{io} = C_{iRO}$ when $a > a_o$

The incipient cavitation coefficients of the plate edge of the Figure 1 corresponds to an upstream pressure of 8 kg/cm^2 abs. and according to the Reference [2] for other different pressure P_1 correct for the term PSE that is the Pressure Scale Effect. Its value is $PSE = 0.53(P_1 - P_v)^{0.3}$ as is reflected in the equation (7).

The Figures 9 and 10 show the values of C_{io} for different sizes of EMA-1 with $d_e/D = 0.2$ and 0.5

3.3 How to design a rotary multi-hole plate (number and size of holes)

The author tested in a Spanish nuclear power plant multi-hole plates with n = 21 holes and d_o = 6.5, 11 and 15 mm in a piping size of D = 100 mm (4 in) and d_o = 13, 22 and 30 mm for D = 200 mm (8 in)

Many other multi-hole plates with different values of n, d_o and D has been designed and installed in several nuclear power and thermal power plants in Spain and abroad, basing in the test results, and all they run according to the design.

For this reason, though theoretically the greater values of n decrease the C_i values, the practical results of the experience don´t show this as evident.

As a suggestion for design, I recommend follow these criteria to define the number of holes n of the multi-hole plate:

For D < 3 in, take n = 4.

For D equal or greater than 3 in and equal or lower than 6 in, take n = 9 or n = 13.

For D > 6 in, take n = 21.

If we start with the known data of D, DP, W, P_1 and P_v, to design the rotary multi-hole plate that pass the flow rate W in kg/h and provokes the pressure drop DP without cavitation, follow these steps:

1. Select a flow velocity of V = 5 m/s to estimate the size D of the rotary multi-hole plate, applying this equation:

$$\mathbf{D = 8.41Q^{0.5}} \qquad (9)$$

2. Calculate the flow coefficient C_{vo} applying the equation (2), taking into account that $Q = v_e W$

   The incipient cavitation coefficient C_{io} must be: $C_{io} < (P_1 - P_v)/DP$
3. Taking into account the Figures 9 and 10 and the value of C_{io} it´s possible to estimate the necessary value of d_e/D.
4. From the Figures 5 and 6 obtain C and $F^{0.5}$.

   Note that to obtain C, first is necessary to calculate the Re number based in the internal diameter of the piping before the rotary multi-hole plate.
5. With the equations (5) and (6) calculate C_v and a. This opening angle is the position must have the rotary multi-hole plate to achieve that the flow rate be Q and the pressure drop DP.

6. Apply the equations (7) and (8) to calculate C_{iV} and C_{iRO} and take as C_{io} the great of them. Compare it with $(P_1 - P_v)/DP$ and if it is lower, the design of the rotary multi-hole plate is correct. If not, select other value of d_e/D or a lower value of D and repeat the process from the step 4.

4. OPERATIONAL EXPERIENCE OF THE ROTARY MULTI-HOLE PLATES. CONCLUSIONS

Two rotary multi-hole plates of D = 100 mm (4 in) with n = 16 holes and d_o = 9 mm are installed, each one, in the minimum flow recirculation lines of the firewater pumps of Almaraz, a Spanish nuclear power plant.

Its incipient cavitation coefficient is very close to 1, they have a pressure drop of more than 9 kg/cm^2 abs. and its backpressure is the atmospheric pressure.

They run completely free of cavitation. In the vertical position with the opening angle a = 0º, they are as multi-hole restriction plates and they have cavitation. When the opening angle is approximately between 3º and 26º with the minimum cavitation coefficient at 20º.

Each device comes from a butterfly valve where the disc were drilled with 16 holes of 9 mm each one.

See the Reference [6] to know more details of these two devices.

These two devices are prototypes that validate the theoretical design of the rotary multi-hole plates for whichever other sizes and the equations showed in this document, so the conclusions are:

1. The rotary multi-hole plates could be used as butterfly control valves with an extremely low cavitation coefficient but they can´t isolate the flow.
2. The rotary multi-hole plates are multi-hole restriction orifice plates that can adjust its position in several opening angles with cavitation coefficients lower than the multi-hole plates.

5. FIGURES

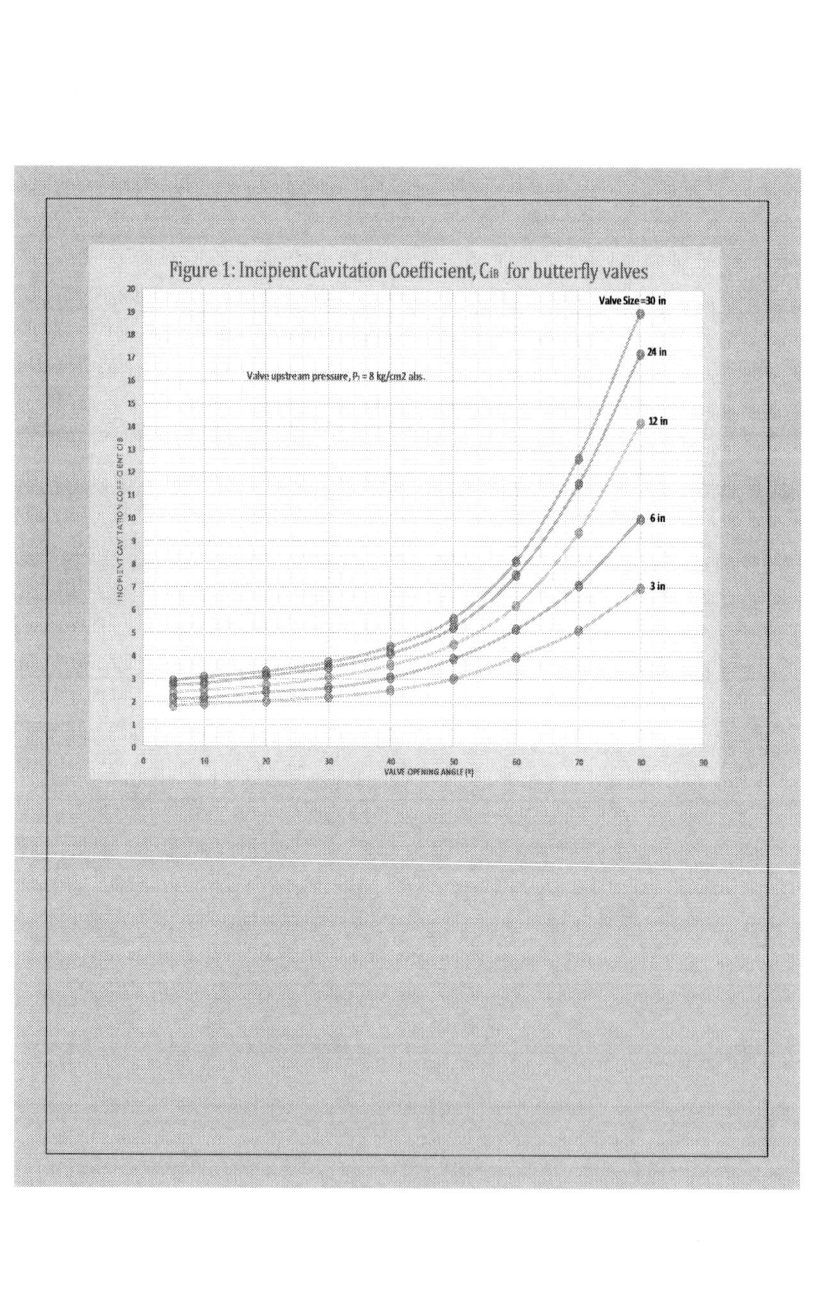
Figure 1: Incipient Cavitation Coefficient, C_{iB} for butterfly valves
Valve Size =30 in
24 in
12 in
6 in
3 in
Valve upstream pressure, P_1 = 8 kg/cm2 abs.
INCIPIENT CAVITATION COEFFICIENT CIB
VALVE OPENING ANGLE (º)
0
10
20
30
40
50
60
70
80
90
1
2
3
4
5
6
7
8
9
10
11
12
13
14
15
16
17
18
19
20

Figure 2: Incipient cavitation coefficient, Ci , for RO multi-hole plates referred to the upstream pressure

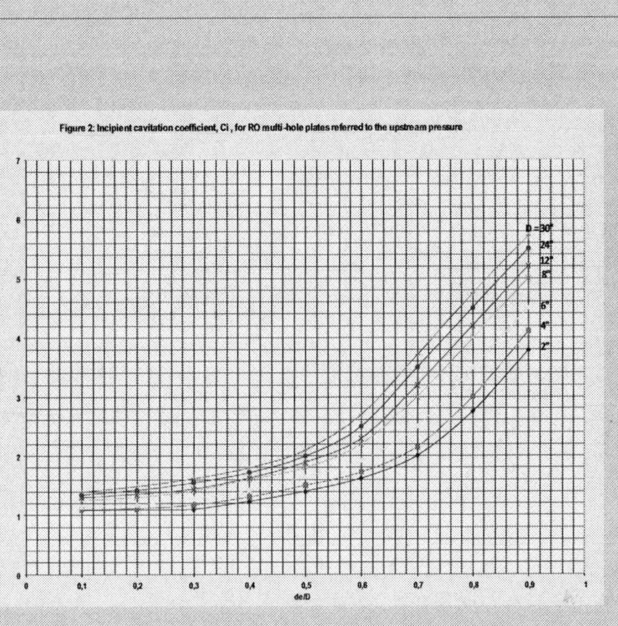

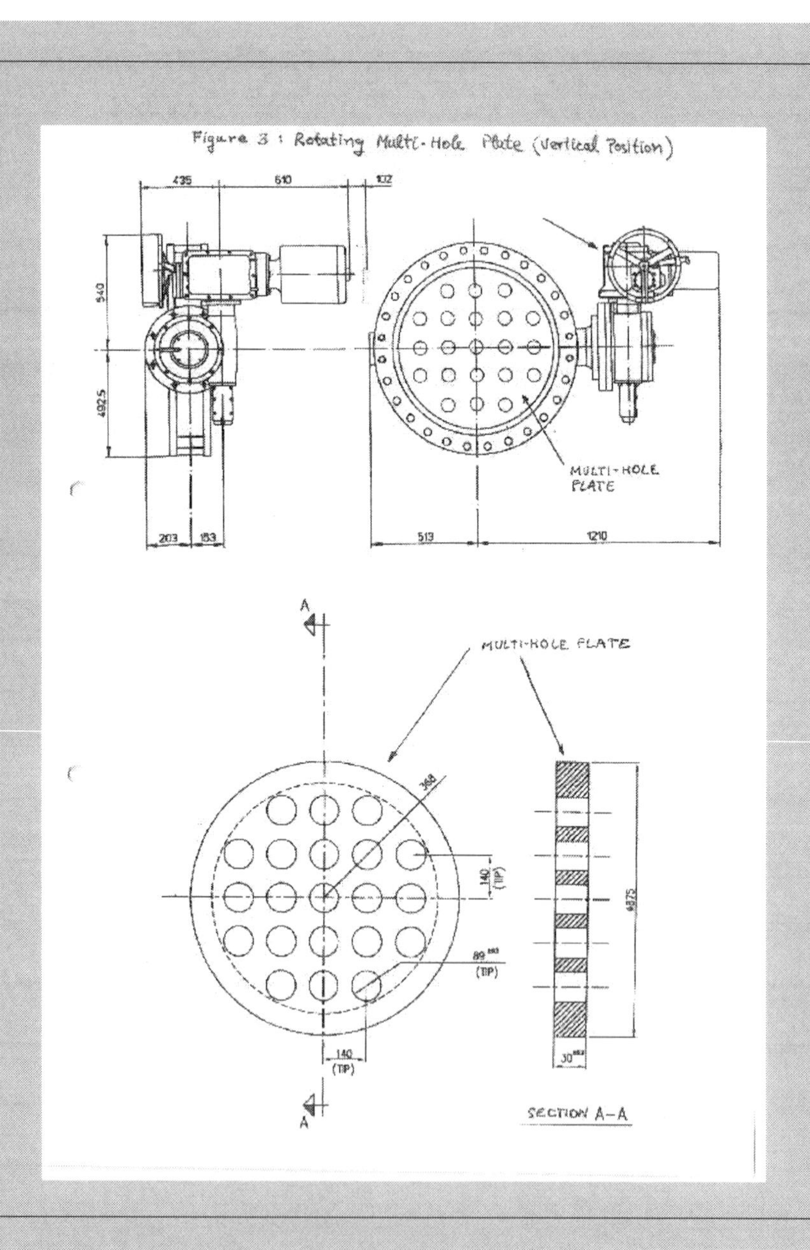

Figure 3 : Rotating Multi-Hole Plate (Vertical Position)

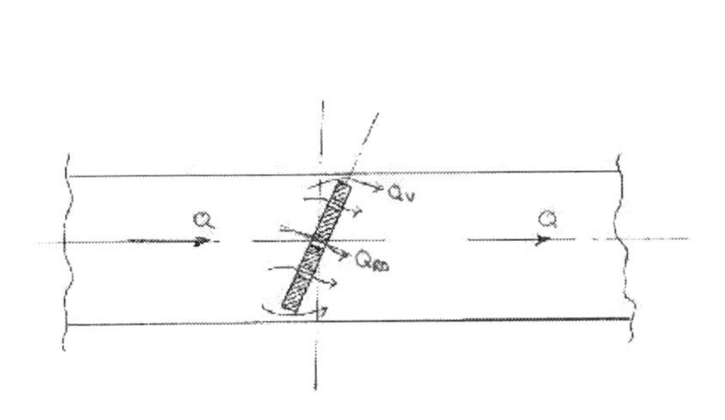

Figure 4 : Rotating Multi-Hole Plate
(Throttling Position)

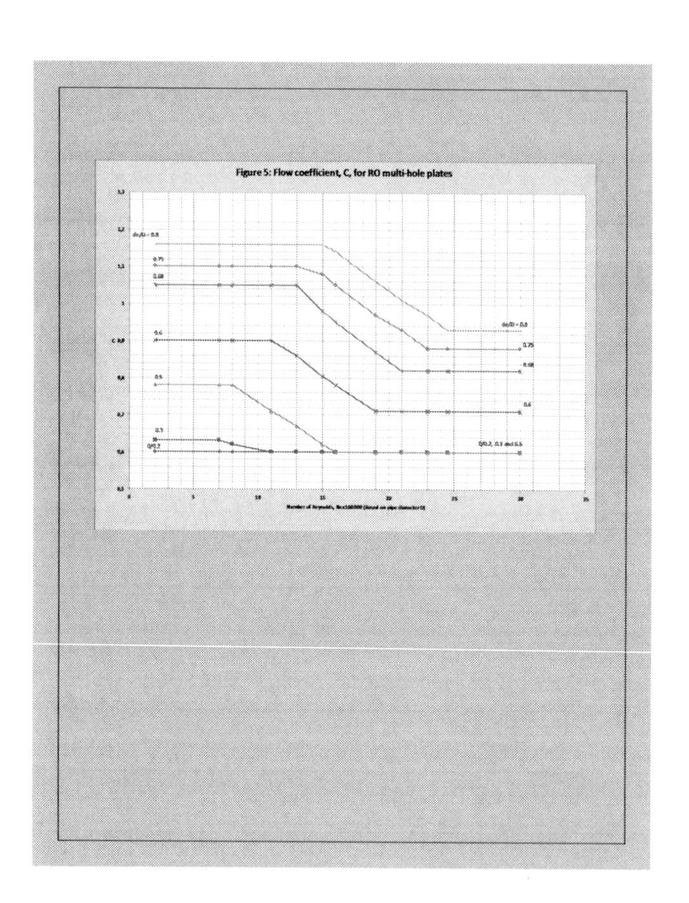
Figure 5: Flow coefficient, C, for RO multi-hole plates
C
0.75
0.6
0.5
0.3
0
5
10
15
20
25
30
35

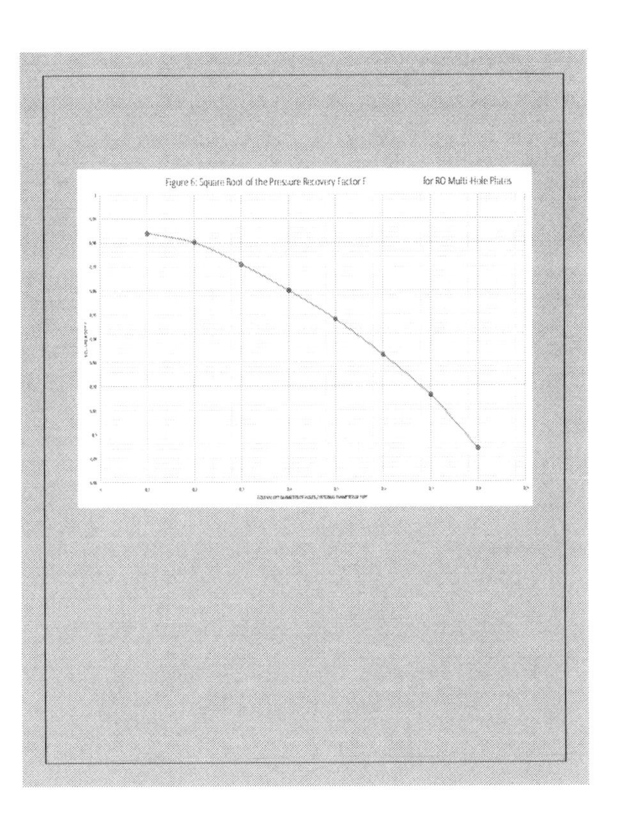
Figure 6: Square Root of the Pressure Recovery Factor F for RO Multi-Hole Plates

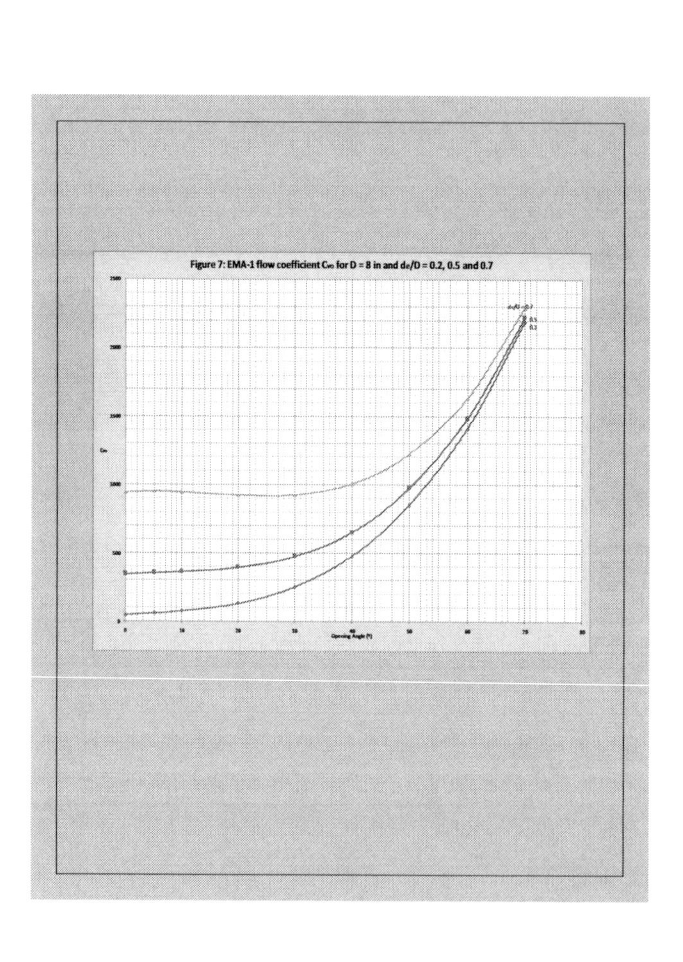

Figure 7: EMA-1 flow coefficient C_{vo} for D = 8 in and d_e/D = 0.2, 0.5 and 0.7

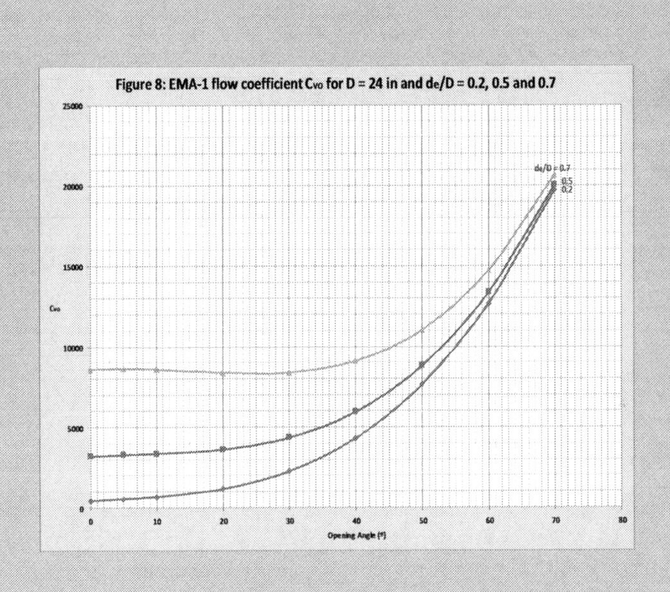

Figure 8: EMA-1 flow coefficient C_{vo} for D = 24 in and d_e/D = 0.2, 0.5 and 0.7

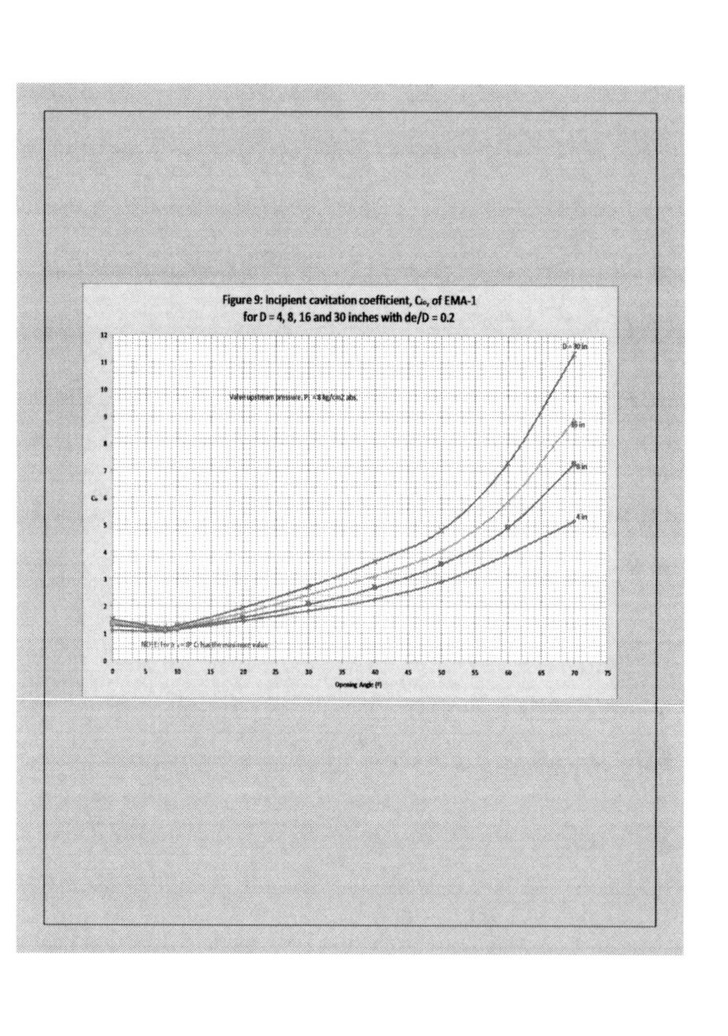
Figure 9: Incipient cavitation coefficient, Cᵢₒ, of EMA-1
for D = 4, 8, 16 and 30 inches with de/D = 0.2
D = 30 in
16 in
8 in
4 in
Opening Angle (°)
0
5
10
15
20
25
30
35
40
45
50
55
60
65
70
75
1
2
3
4
5
6
7
8
9
10
11
12

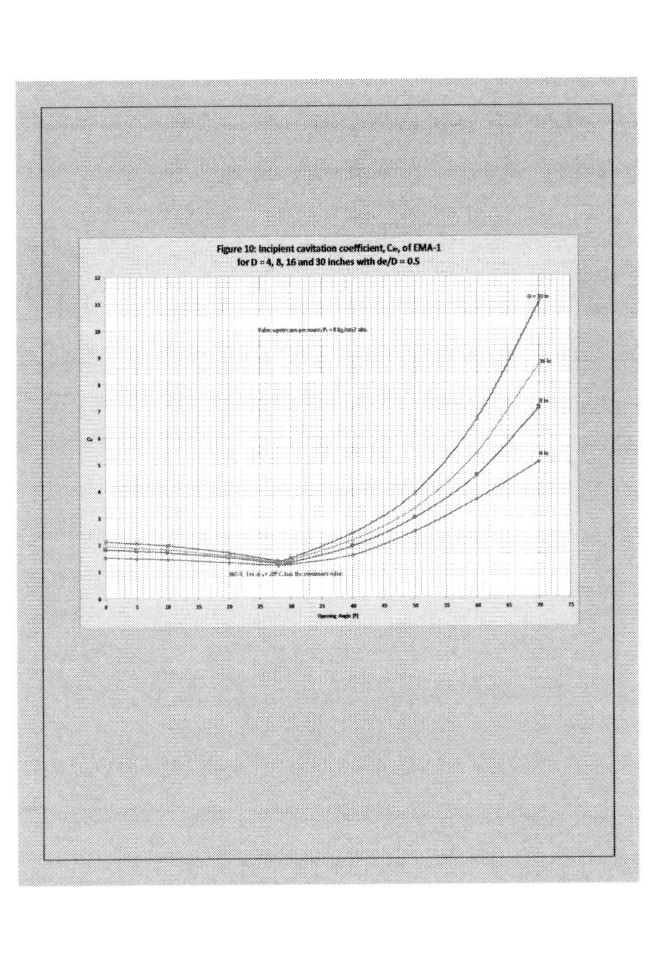

Figure 10: Incipient cavitation coefficient, C_{ip}, of EMA-1 for D = 4, 8, 16 and 30 inches with de/D = 0.5

6. REFERENCES

[1] E. Casado. "Avoid cavitation in butterfly valves". HYDROCARBON PROCESSING. August 2006.

[2] J. P. Tullis. "Cavitation Guide for Control Valves". NUREG/CR-6301 NRC FIN L 2574. April 1993.

[3] E. Casado. "Look at orifice plates to cut piping noise, cavitation". POWER. September 1991.

[4] E. Casado. "Eliminación de la vibración, ruido y cavitación en las tuberías mediante placas multiperforadas". INGENIERIA QUIMICA. Diciembre 1991.

[5] E. Casado. "Experiencia del comportamiento de las placas multiperforadas". INGENIERIA QUIMICA. Octubre 2003.

[6] E. Casado. "Eliminate cavitation in your piping systems". HYDROCARBON PROCESSING. February 2012.

[7] Crane Technical Paper Nº 410. "Flow of Fluids Through Valves, Fittings and Pipe".

7. NOMENCLATURE

D, d_e = Disc plate diameter and holes equivalent diameter (mm)

P = Pressures (kg/cm^2 abs.)

DP = Pressure drop (kg/cm^2)

Q = Volumetric flow rate (m^3/kg)

CHAPTER 7

ONE HOLE AND MULTI-HOLE THICK ORIFICE PLATES

DESIGN GUIDE OF THICK ORIFICE PLATES

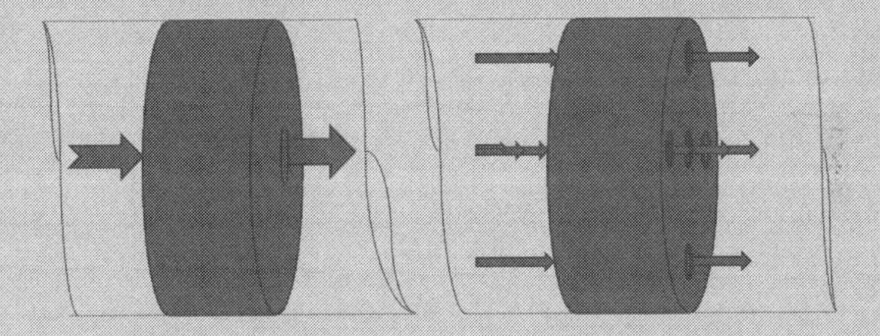

(One Hole Thick Plates and Multi-Hole Thick Plates)

Emilio Casado

Consulting Engineer

Revision 2

August 2021

CONTENT

INTRODUCTION

The thick orifice plates of one central hole and multi-holes are similar to the thin orifice plates. They are intermediate devices between the thin plate restriction orifices and the long-thick restriction orifices. See the References [2], [4] and [5].

In the design of a thin plate restriction orifice, if the pressure drop is high, the necessary plate thickness may result big and the thin plate must change to a thick plate.

There is no unanimity among the authors how to classify the plates between thin and thick. Some of them consider that a plate is thick when $t/d_o > 1$, being t, the plate thickness and d_o, the hole diameter.

Conservatively in this Guide, I consider the following classification:

Thin orifice plates: $t/d_o < 1$

Thick orifice plates: t/d_o equal or greater than 1 and less than 7.

Long-thick orifices: t/d_o equal or greater than 7.

See the Reference [11].

NOTE: The thickness t of the long-thick orifices is its length L.

The hydraulic characteristic differences among the thin orifice plates, the thick orifice plates and the long-thick orifices are:

The thin orifice plates provoke that the fluid vena contract (the fluid stream has at the vena contract section a diameter less than the hole diameter) be formed downstream the plate. For this reason, the thin plates when the hole is sharp-edged do not provoke critical conditions in the fluid. That is, the fluid, when it is compressible, do not choke.

The thick orifices have the vena contract inside the hole, and the fluid may reattach the wall of the hole before exiting the plate. For compressible fluids, the thick plates may have critical conditions at the exit of the hole. For liquids, the thick orifices may have or not, cavitation. When they have cavitation, it reaches the exit plate and it is super-cavitation because the thickness is not enough to maintain the cavitation inside the hole.

This Revision 2, gives more appropriated values of the expansion factor Y according to the new Reference [11].

The long-thick orifices may have critical conditions at its exit, may have or not cavitation but not super-cavitation because its length L is enough to maintain the cavitation inside it.

This Guide exposes how to design the thick restriction orifice plates of one hole and multi-hole for liquids and compressible fluids with t/d_o equal or greater than 2 and equal or less than 5, with more than 3 holes in the case of the multi-hole plates, as shown the Figures 1 and 2.

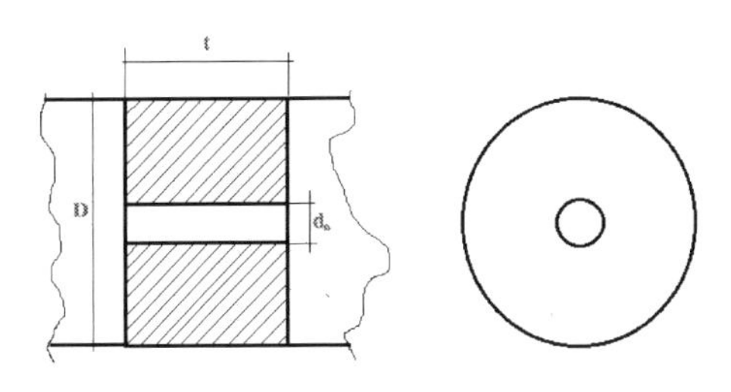

Figure 1: Thick orifice plate with 1 hole

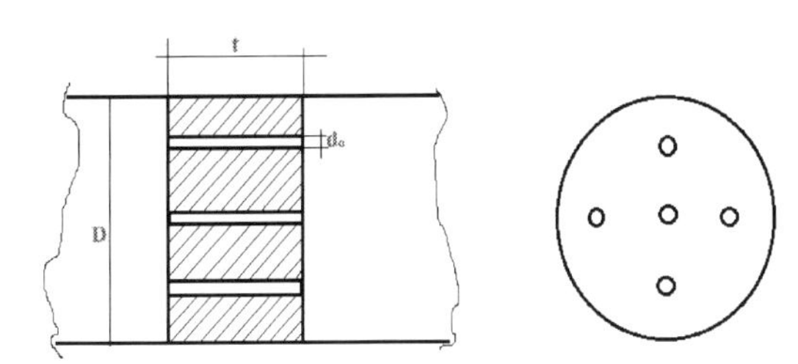

Figure 2: Multi-Hole Thick Orifice Plate with n = 5 holes

DESIGN OF THICK ORIFICE PLATES WITH ONE CENTRAL HOLE

Design for liquids. The flow coefficient C and the incipient cavitation coefficient S_i

The relation between the pressure drop in the plate and the flow rate through it in the case of liquids is:

$$Q = 1.25Cd_o^2[v_e(P_1 - P_2)]^{0.5} \quad (1)$$

Q = Flow rate (m^3/h)

P_1= Pressure of the fluid before the plate (kg/cm^2 abs.)

P_2 = " " " " after " " (kg/cm^2 abs.)

v_e = Specific volume of the fluid (m^3/kg)

d_o = Diameter of the hole (mm)

C = Flow coefficient of the plate

The Reference [3] recommends use the following equation to estimate C when there is no cavitation:

$$C = (0.827 - 0.0085t/d_o)/[1 - (d_o/D)^4]^{0.5} \quad (2)$$

The values that give this equation are similar to those corresponding to the long-thick orifices of the Reference [4]. In this Reference for a long-thick orifice with a sharp-edge entrance and relative roughness of $e/d_o = 0.001$ is $C = 0.82$ for $t/d_o = 5$ and $C = 0.835$ for $t/d_o = 2$. So, use the equation (2) to calculate C for the thick plates with one hole.

The Figure 3 is the graph of C.

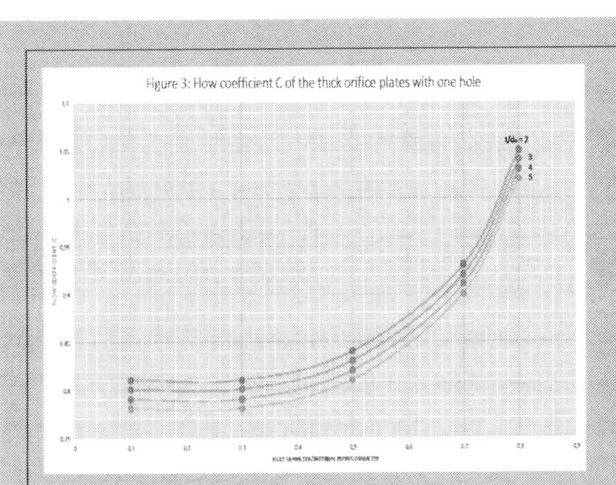

Figure 3: Flow coefficient C of the thick orifice plates with one hole

NOTE: This graph corresponds to the values of C when the thick orifice plate has not cavitation.

When the plate has cavitation, the References [3] and [4] say that C is:

$$C = 0.615[(P_1 - P_v)/(P_1 - P_2)]^{0.5} \qquad (3)$$

There is cavitation if $(P_1 - P_v)/(P_1 - P_2) < S_i$

S_i is the incipient cavitation coefficient and P_v is the vapor pressure of the fluid in kg/cm^2abs.

For the thick plates with one hole of sharp edge entrance and relative roughness of e/d_o = 0.001, consider according to the Reference [4], that S_i is 1.81 for $t/d_o = 5$ and $S_i = 1.9$ for $t/d_o = 2$, with an average value of **$S_i = 1.85$** for whichever be d_o/D. For greater values of e/d_o S_i will be lower.

The Reference [3] gives $S_i = 1.82$ for $t/d_o = 2$, $d_o/D = 0.22$ and D = 1 inch. This Reference states that for other t/d_o and d_o/D values, it is necessary investigate in future works.

Take as a base reference $S_i = 1.85$ for $2 \leq t/d_o \leq 5$, D = 1 inch $d_o/D = 0.22$ and compare this value with the incipient cavitation coefficient C_i of the thin plate with one hole, D = 1 inch and $d_o/D = 0.22$ that is $C_i = 1.65$, the relation between both is:

$S_i/C_i = 1.85/1.65 = 1.12$

Taking the values of C_i from the Reference [11] and multiplying them by 1.12 the result are the estimated theoretical values of S_i of the thick orifice plates with one hole, which correspond to the Figure 3a.

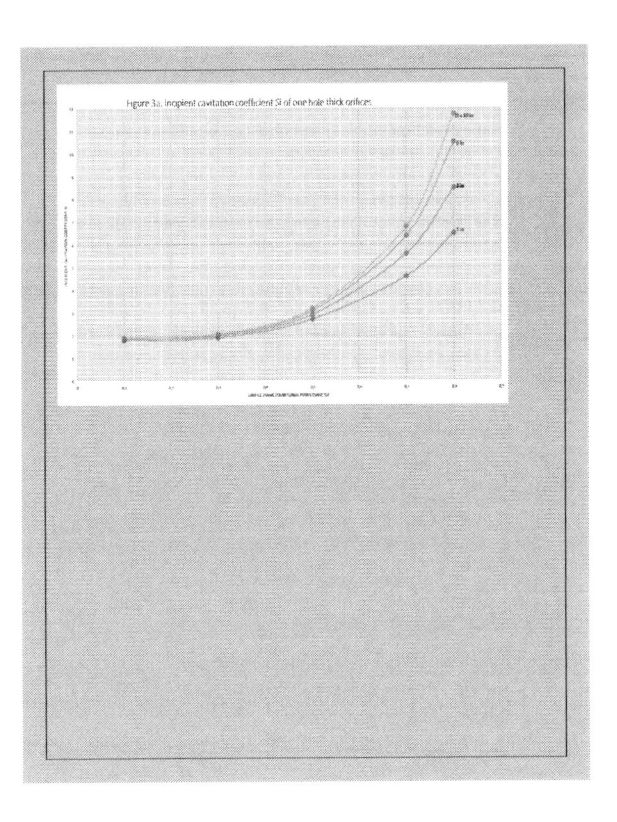

Figure 3a. Incipient cavitation coefficient Si of one hole thick orifices

When the thick plate has cavitation, the flow rate is constant and as in the long-thick orifices, Q is:

$$\mathbf{Q = 0.764d_o^2[v_e(P_1 - P_v)]^{0.5}} \qquad (4)$$

In conclusion, for liquids, the thick orifice plates with one hole, have the flow coefficients C given by the Figure 3, when there is no cavitation or by the equation (3) when there is cavitation. The incipient cavitation coefficients S_i correspond to the Figure 3a, for the range $2 \leq t/d_o \leq 5$.
The difference between the one-hole orifice plates and the long-thick orifices is that in the thick plates, the cavitation reaches its exit (super-cavitation) and in the long-thick orifices, do not.

Design for steam and gases. The flow coefficient C and the expansion factor Y

When the fluid is steam or a gas, the relation between the flow rate and the pressure drop in the plate is:

$$Q = 1.25Cd_o^2Y[v_e(P_1 - P_2)]^{0.5} \quad (5)$$

This equation is the same as the equation (1) but including the expansion factor Y of the Figures 4 and 5. In these Figures the coefficient a is the relation between the specific heats at constant pressure and volume. Also, in the Figures 9 and 10.

The flow coefficient C is the same of the Figure 3 and the values of Y are intermediate between those corresponding to the thin orifice plates with one hole (See the Reference [2]) and those corresponding to the long-thick orifices of the Reference [5].

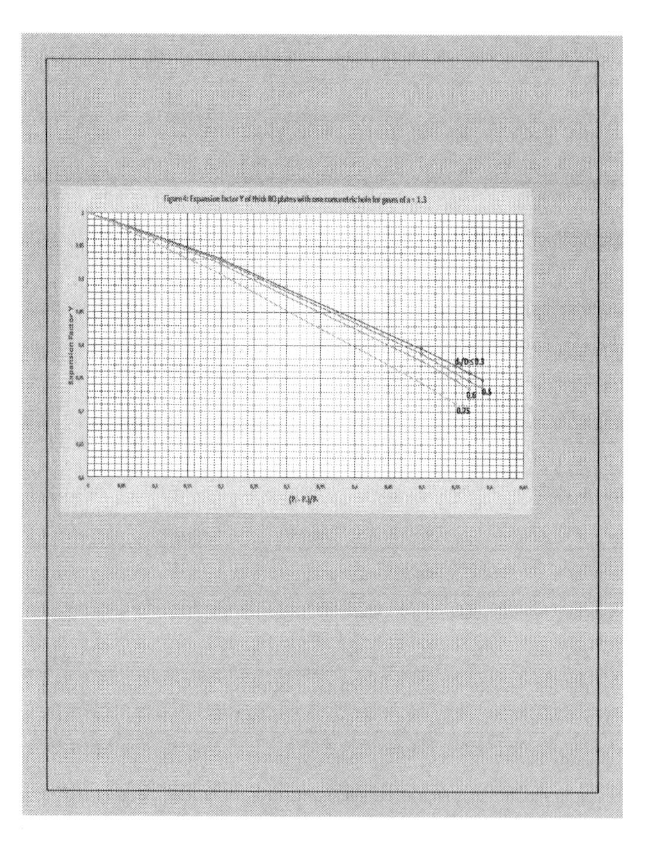

Figure 4: Expansion factor Y of thick RO plates with one concentric hole for gases of κ = 1.3
Expansion Factor Y
d/D ≤ 0.3
0.5
0.6
0.75

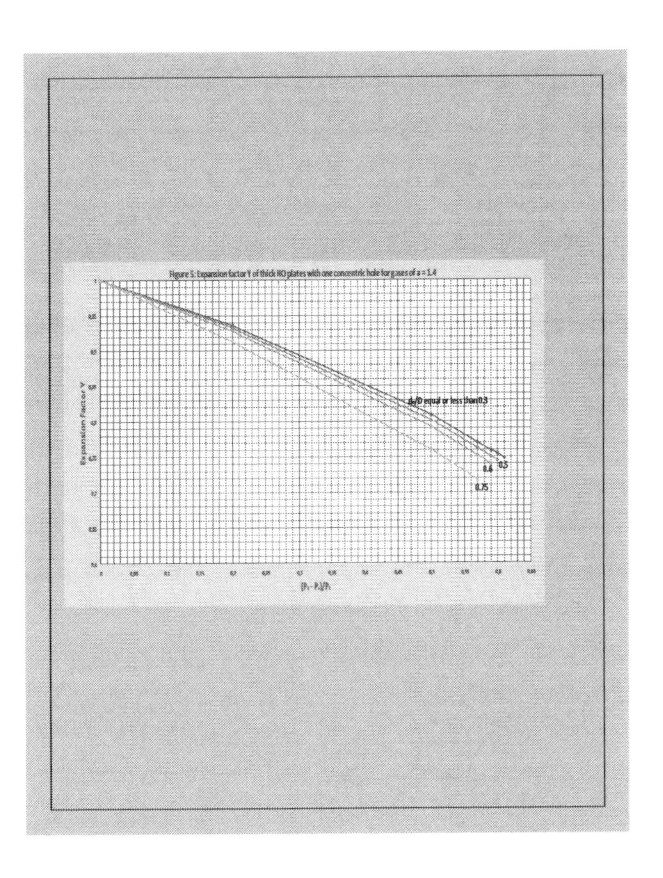
Figure 5: Expansion factor Y of thick RO plates with one concentric hole for gases of a = 1.4
Expansion Factor Y
d_0/D equal or less than 0.3
0.5
0.6
0.75
$(P_1 - P_2)/P_1$

Design for saturated water. The flow coefficient C and the expansion coefficient C_E

When the fluid is saturated water, the relation between the flow rate and the pressure drop in the thick plate is:

$$Q = 0.19C_E(2.65 + \ln P_s)Cd_o^2[v_e(P_s - P_2)]^{0.5} \quad (6)$$

This equation comes from the Reference [5] because the resistance coefficient $K = 1/C^2$, that is deduced taking into account the equation (1) and the equation applied to a pipe with the same length t and the same diameter as the orifice hole d_o, that provokes the same pressure drop for the same flow. This equivalent equation is: $Q = 1.25d_o^2[v_e(P_1 - P_2)/K]^{0.5}$

When $P_1 = P_s$ the relation between the pressure drop and the flow rate of saturated water through a pipe with the diameter d_o and the resistance coefficient K is $Q = 0.19C_E(2.65 + \ln P_s)d_o^2[v_e(P_s - P_2)/K]^{0.5}$

Q = Flow rate (m^3/h)

C_E = Expansion coefficient of the saturated water (Figure 6)

P_s = Saturation pressure of the water before the plate (kg/cm^2 abs.)

C = Flow coefficient of the plate (Figure 3)

d_o = Diameter of the hole (mm)

v_e = Specific volume of the water before the plate (m^3/kg)

P_2 = Pressure of the water after the plate (kg/cm^2 abs.)

The flow coefficient C is of the Figure 3.

The expansion coefficient C_E of the saturated water for the flow through pipes is included in the Reference [5] for the long-thick orifices.
As $K = 1/C^2$ and the minimum and maximum values of C, applying the equation (2) are

respectively 0.784 and 1.05 for t/d_o between 2 and 5, the maximum and minimum values of K are respectively 0.9 and 1.63.

The Figure 3 of the Reference [5] corresponds to the graph of C_E developed by the author. From this Figure, extrapolating the values of C_E until K = 0.2, it is possible to deduce the graph of the Figure 6. This graph corresponds to the estimated expansion coefficient C_E of the thick orifice plates for the flow of saturated water through them.

From the Critical Conditions Line of this graph, it is possible to deduce the relation between the saturation pressure of the water at the entrance of the thick plate and the critical pressure at the exit. This relation with a deviation less than 2 % corresponds to the equation:

$$\mathbf{P_{CR} = 0.705C^{0.272}P_s} \qquad (7)$$

For example, a thick orifice plate with $d_o/D = 0.5$ and $t/d_o = 4$, according to the Figure 3 has C = 0.82 and if P_s = 100 kg/cm^2 abs., P_{CR} = 66.8 kg/cm^{2}abs.

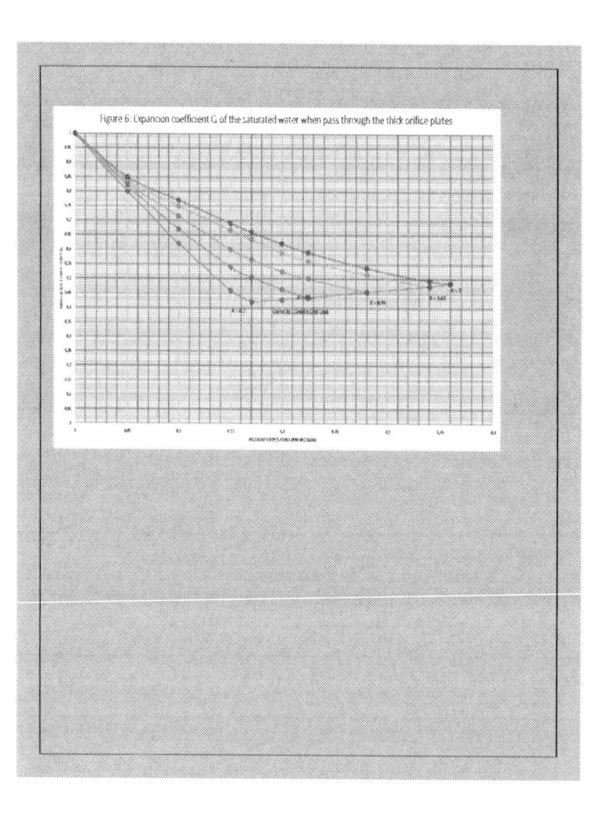
Figure 6: Expancion coefficient G of the saturated water when pass through the thick orifice plates

DESIGN OF MULTI-HOLE THICK PLATES

Comparing the one-hole **thin** plates with the multi-hole **thin** plates, the second have similar or greater flow coefficients C and lower incipient cavitation coefficients, so we can assume that in the thick plates will be similar.

Design for liquids. The flow coefficient C and the incipient cavitation coefficient S_i

Take the Reference [7] to establish the flow coefficients C when there is no cavitation. In these Reference, the values of C obtained in the tests are independent of the Reynolds number, that is, independent of the liquid hole velocity.

The thick tested plates in the range of t/d_o between 2 and 5 had t/d_o equal to 3.125 and 3.375, that is approximately equal to 3.

The Figure 7 shows the values of C as a function of d_e/D and t/d_o, being d_e the equivalent diameter of the n holes. If d_o is the diameter of every hole, d_e is: $d_e = d_o n^{0.5}$

The relation among the values of C for the same d_e/D and different t/d_o is the same as in the Figure 3 for one-hole plates.

NOTE: The Reference [7] gives the flow resistance coefficients K obtained in the tests and the values of C correspond to $C = 1/K^{0.5}$

When there is cavitation, calculate the values of C applying the equation (3) as in the case of the thick orifice plates with one hole and calculate the flow rate Q with the equation (4) changing d_o by d_e

Comparing the values of C of the multi-hole thick plates in the Figure 7 with those of the one-hole thick orifice plates in the Figure 3, we can see that are greater. It is the same as in the thin plates.

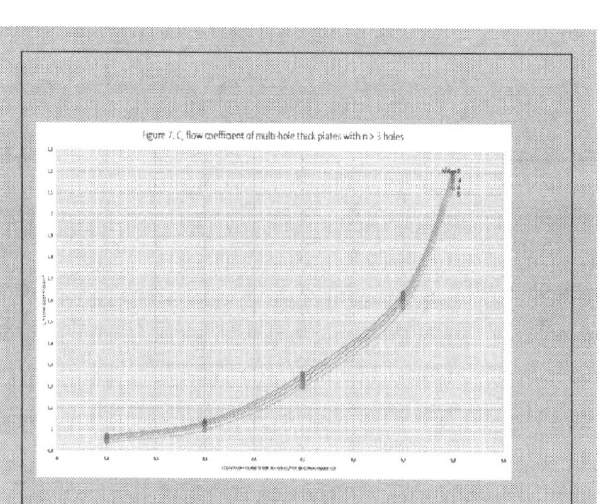

Figure 7. C, flow coefficient of multi-hole thick plates with n > 3 holes

NOTE: This graph corresponds to the values of C when the multi-hole thick orifice plate has not cavitation.

When the plate has cavitation, apply the equation (3) to calculate C.

The plate will have cavitation if $(P_1 - P_v)/(P_1 - P_2) < S_i$

The Reference [7] gives some values of S_i, but they seems very conservative because in its Figure 12 includes also the values of S_i for thin plates ($t/d_o < 2$) and they are much higher than those published in the Reference [8].

For this reason, I prefer to deduce the incipient cavitation coefficients of the multi-hole thick plates based on those of the on- hole thick plates of the Figure 3a, corrected with the reduction coefficient that have the multi-hole thin plates, with respect to the one-hole thin plates.

The result is the graph of the Figure 8.

When the thick plate has cavitation, apply the equation (4) to calculate the flow rate.

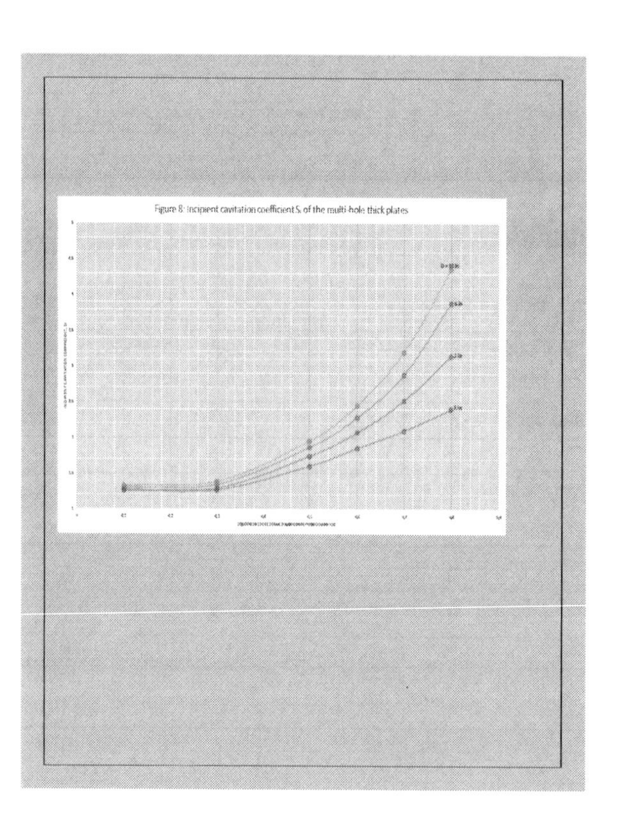

Figure 8: Incipient cavitation coefficient S_i of the multi-hole thick plates

Design for steam and gases. The flow coefficient C and the expansion factor Y

To calculate the relation between the flow rate Q and the pressure drop in the plate, $P_1 - P_2$, use the equation (5), changing d_o by d_e, this is:

$$Q = 1.25 C d_e^2 Y [v_e (P_1 - P_2)]^{0.5}$$

Use also the Figure 7 to obtain C and the Figures 9 and 10 to obtain Y.

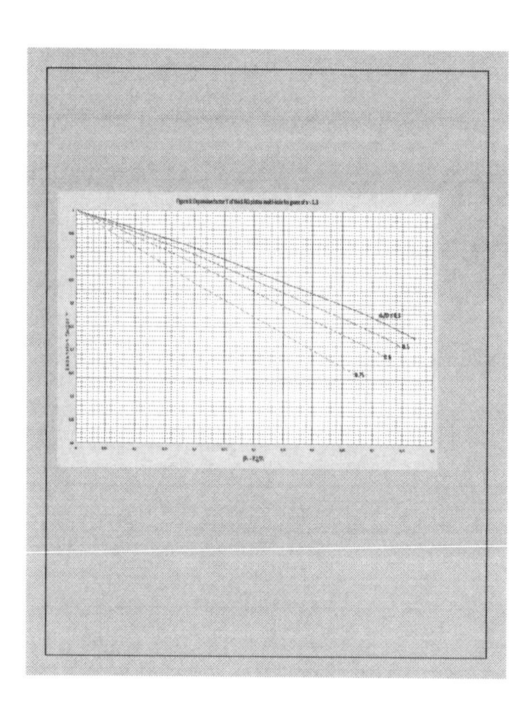

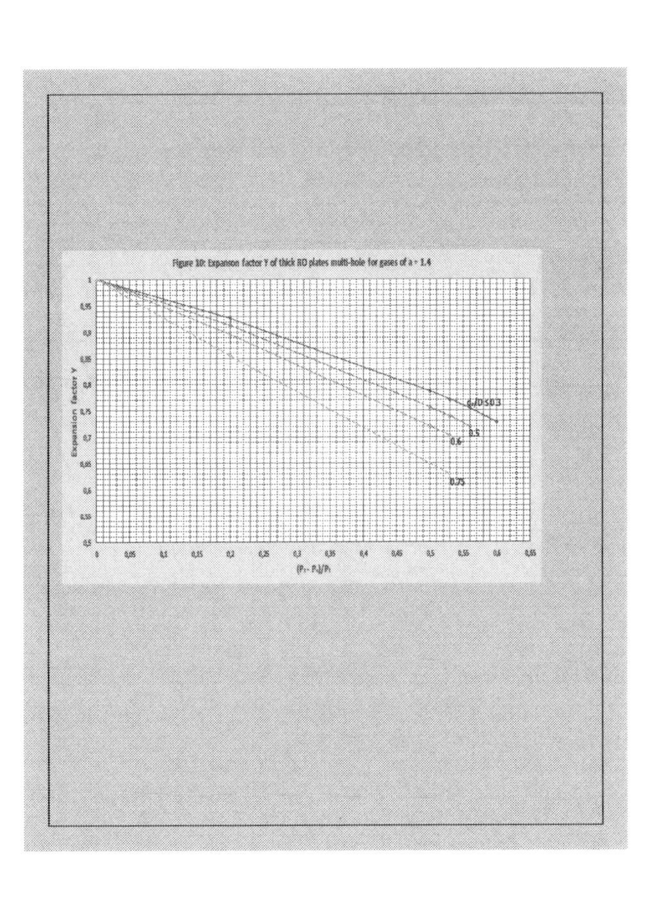

Figure 10: Expanson factor Y of thick RO plates multi-hole for gases of a = 1.4

Design for saturated water. The flow coefficient C and the expansion coefficient C_E

To calculate the relation between the flow rate and the pressure drop through the plate, use also the equation (6) changing d_o by d_e, that is:

$$Q = 0.19C_E(2.65 + \ln P_s)Cd_e^2[v_e(P_s - P_2)]^{0.5}$$

Use the Figure 7 to obtain C and the Figure 6 to obtain C_E

Use also the equation (7) to calculate the relation between the saturation pressure at the plate inlet and the critical pressure at the plate outlet.

STRUCTURAL DESIGN

The one-hole thick plates have the same structural design as the one-hole thin plates of the Reference [2]. The Chapter 1 of the DESIGN HANDBOOK OF RESTRICTION ORIFICES, included as Reference [12] exposes also how to develop the calculations.

The multi-hole thick plates have the same structural design as the multi-hole thin plates of the Reference [10]. Also, the Chapter 2 of the Reference [12] exposes how to develop the calculations.

REFERENCES

[1] Crane Technical Paper Nº 410. "Flow of Fluids Through Valves, Fittings and Pipe".

[2] E. Casado. "Design guide for restriction orifice plates with one central hole". Revision 3, June 2018. Research Gate.

[3] B. Ebrahimi et all. "Characterization of high-pressure cavitating flow through a thick orifice plate in a pipe of constant cross section". International Journal of Thermal Sciences 114 (2017) 229-240.

[4] E. Casado. "Long and thick restriction orifices for liquids". Revision 2, June 2018. Research Gate.

[5] E. Casado. "Long and thick restriction orifices for saturated water, steam (saturated and reheated) and gases.

[6] F. Self and S. Ganjam (Bechtel); G. Jacobs (Virtual Materials Group). "Models for multi-phase & single-phase flow in pressure relieving system using Bernoulli integration". American Institute of Chemical Engineers, 2015 Spring Meeting.

[7] D. Maynes; G. J. Holt and J. Blotter. "Cavitation inception and head loss due to liquid flow through perforated plates of varying thickness". Journal of Fluids Engineering. March 2013 Vol. 135/031302-1- 031302-11.

[8] E. Casado. "Look at orifice plates to cut piping noise, cavitation". Power, September 1991.

[9] E. Casado. "Improve pressure drop calculations for saturated water lines". Hydrocarbon Processing. August 2013.

[10] E. Casado. "Design guide for multi-hole restriction orifice plates with n > 3 holes. Research Gate.

[11] E. Casado. "Design of thin and thick RO plates for the flow of gases and reheated steam". Amazon and Research Gate. July 2021.

[12] E. Casado. "Design Handbook of Restriction Orifices". Research Gate and Amazon. August 2021.

Printed in Poland
by Amazon Fulfillment
Poland Sp. z o.o., Wrocław
10 November 2023

3d4bda07-14bd-4fd4-baa6-e97b52230a01R02